THE GENERAL HISTORY OF ASTRONOMY

Volume 4

Astrophysics and twentieth-century astronomy to 1950: Part A

Published under the auspices of the International Astronomical Union
and the
International Union for the History and Philosophy of Science

THE GENERAL HISTORY OF ASTRONOMY

General editor: Michael Hoskin, University of Cambridge

Volume 4

Astrophysics and twentieth-century astronomy to 1950: Part A

EDITED BY

OWEN GINGERICH

Harvard-Smithsonian Center for Astrophysics

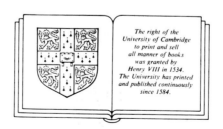

The right of the
University of Cambridge
to print and sell
all manner of books
was granted by
Henry VIII in 1534.
The University has printed
and published continuously
since 1584.

CAMBRIDGE UNIVERSITY PRESS

Cambridge

London New York New Rochelle

Melbourne Sydney

Published by the Press Syndicate of the University of Cambridge
The Pitt Building, Trumpington Street, Cambridge CB2 1RP
32 East 57th Street, New York, NY 10022, USA
296 Beaconsfield Parade, Middle Park, Melbourne 3206, Australia

First published 1984

Printed in Great Britain at the University Press, Cambridge

Library of Congress catalogue card number: 83-10164

British Library cataloguing in publication data

Astrophysics and twentieth century astronomy to
1950 – (The general history of astronomy; v.4)
Pt A
1. Astrophysics – History
I. Gingerich, Owen II. Series
523.01   QB461

ISBN 0 521 24256 8

CONTENTS

THE CONTENTS OF PART B OF VOLUME 4 ARE:

FOREWORD

Historians often remark that such-and-such is 'known'. By this they mean that a colleague in some part of the world has investigated the matter and published a well-documented account of his investigation that has convinced his fellow-specialists. The purpose of the *General History of Astronomy* is to take what is 'known' in this restricted sense and make it known in the common sense of the word; that is, readily available to a wide readership. With the blessing of the International Astronomical Union and the International Union for History and Philosophy of Science, historians of astronomy from every part of the world have pooled their expertise to present an account for the non-specialist of the development of astronomy from earliest times, often making available to a wider readership insights that have hitherto been 'known' to a select few.

Some historians of science never write for a wider readership, and to them it has become second nature to assume an audience of sceptical colleagues whom they must convince by citing references for every item of evidence. Such historians will be disconcerted to find few references in the present work. The reason is that here our authors are presenting to a wider public research that for the most part has already been documented in the professional literature, and to fill our pages with references that those interested can readily find elsewhere would obscure our purpose. Instead we give a realistic bibliography of Further Reading, of works written in the major languages and accessible in good libraries; those who wish to explore further will find in these works an introduction to the more specialist literature.

When this enterprise was first proposed, some historians of astronomy questioned whether our profession had yet reached the stage when such a synthesis should be attempted. Some areas have been thoroughly researched, but others have been barely reconnoitred. Our authors have been encouraged to say frankly when they are reporting provisional judgements or even first impressions. No one has been more aware than the editors, of the limitations of what is currently 'known'. We must be content if these volumes represent a substantial advance on what has been available hitherto.

Michael Hoskin
General editor

PREFACE

Whether the historian accepts an evolutionary or a revolutionary model for the development of scientific understanding, he will surely recognize in the history of astronomy certain periods of intensive innovation. We have chosen two of these to set the limits on Volume 4 of this series.

In the 1850s J.B.L. Foucault discovered the technique of depositing silver on glass, G.P. Bond and others used the newly-developed collodion plates to photograph stars, and G.R. Kirchhoff together with R.W.E. Bunsen established the foundations of spectral analysis. The first two of these innovations made their impact on astronomy rather slowly: not until the invention of the dry plate around 1880 did photography begin its profound alteration of observational techniques in astronomy, and not until the end of the century did the silvered-glass reflector become the instrument of choice for studies of sidereal structure and stellar evolution. However, the discoveries of Kirchhoff and Bunsen, though not without significant antecedents in the studies of light and colour, immediately opened up an entire new arena of scientific investigation. Indeed, the entire balance of research effort was redirected from investigations of the positions and motions of astronomical bodies to studies of their physical composition. Thus 1859, the year of Kirchhoff's initial paper on spectral analysis, saw the birth of astrophysics, or perhaps more accurately, the birth of cosmochemistry. It is for these reasons that we choose the decade of the 1850s as the starting point for the volume on modern astronomy.

The choice of the early 1950s as the final cut-off is more controversial. It was the time following World War II when many of the earlier approaches to astronomical topics found their natural fulfilment. By 1949 the first discrete radio sources had been discovered and identified; by 1951 the con-

cept of stellar populations had been elaborated and the distance scale of the universe revised, and the spiral arms of the Milky Way had been detected both by conventional optical means based on the new ideas of stellar populations and by the 21-centimetre line of radio astronomy. In 1952 the triple-alpha process had joined the earlier carbon cycle in the explanation of stellar energy sources, and semi-empirical calculations of the evolution of red giants had opened the door to understanding the life-cycles of the stars. Nevertheless, the burst of new technology – computers, radio astronomy, space probes – had not yet changed the face of astronomy. Hence the early 1950s can be seen as the preface to an explosive growth in our physical understanding of the cosmos, a growth made possible by a veritable electronic revolution of new computing devices and new data acquisition techniques. Consequently, the middle of the twentieth century provides a particularly suitable break point in our account of the history of astronomy.

At issue, however, is whether the *General History* ought to stop there, or whether it should also attempt to delineate the magnificent achievements of the past three decades. We have chosen to end both this volume and the series around mid-century because, for the post-1950 period, a perspective for picking out the most significant and interesting themes from the immense clutter of detail is as yet difficult to acquire. We are far more apt to be biased by incomplete knowledge and individual prejudices. Moreover, had we attempted to continue beyond this date, we would have encountered, from eager and expectant living astronomers, the thorny problem of defending our particular selection of materials!

Our conception of the nature of this collaborative enterprise is closely related to our decision to bring the volume to an end at the midpoint of the twentieth century. We have not attempted to compile a compendium of astronomy with historical references in the style of Rudolf Wolf's *Handbuch der Astronomie* (Zurich, 1890). A series of names and dates may be accurate and comprehensive, but they do not constitute history. We must distinguish between facts concerning the past and historically significant facts: historically significant facts are those used by historians for illuminating the process of creation and discovery. Consequently, the lack of any particular entry in our index must not be interpreted as a denial of the importance of the name or reference for astronomy today. But from an historical point of view it is sterile to weave in every name or reference in an encyclopaedic fashion, useful as this may be in other contexts.

Our view of the writing of history of science is also quite different from the compilation of a review that attempts to tell us how things are in the universe, with historical references primarily to those "who got it right".

We believe that *the prime duty of the historian of astronomy is to illuminate his science as a creative human activity of the astronomical community of the time.* He must seek to enlarge his reader's experience by transporting him to another age, helping him see the problems as they then appeared. He must set in context the theories that had been inherited from the immediate past, the anomalies thrown up by the evidence of the time (genuine or illusory), the instrumentation or technical apparatus available in that epoch, the contacts between workers in the field, and the communication (or the lack thereof) with those outside the astronomical community. We write *history* when we enable our readers to see sympathetically the problems facing our predecessors in the fitful, sometimes sideways, development of scientific knowledge.

Within this framework, we have not attempted to establish 'firsts' or to set priorities. The first speculative occurrence of an idea is generally far less significant than its later emergence, possibly in other hands, supported by persuasive arguments; for example, A. S. Eddington's assertions that stellar luminosity derived from nuclear energy, made around 1920, are far more noteworthy than James Jeans's embryonic suggestion of 1904. On the other hand, the earliest, faltering steps in the understanding of some astronomical phenomenon or physical process may be more revealing than the final clarifying stages of a problem. Indeed, there may be as much or more to be learned about human creativity by probing an approach that ultimately turns out to be unacceptable than by concentrating on the pathway to the 'correct' solution. Nevertheless, we necessarily write from the clarifying perspective of hindsight, and although we must always be on guard to prevent this viewpoint from distorting the historical picture, we take advantage of it in selecting the most significant threads for weaving our tapestry.

Within our chosen time span and in conformity with the historiographical principles outlined above, we have attempted to select the major themes in the development of astronomy, and to find able authors to articulate these themes for us. We have also sought the advice of numerous referees and critics, who have helped improve the balance and accuracy of these presentations. Our long-suffering authors have had to endure a considerable amount of editorial reworking from both the volume editor and the general editor, so as to guarantee the dovetailing and a reasonable level of homogeneity of the sometimes disparate entries. We are grateful to them, to the referees and to the editors of Cambridge University Press for their cooperation in making this volume possible.

Cambridge, Massachusetts,   Owen Gingerich
December, 1982

ACKNOWLEDGEMENTS

Although the general scope of this volume was already delineated at an editorial board meeting held in Cracow in August, 1972, the specific form evolved only through extensive discussions in which Joseph Ashbrook, David Dewhirst, Michael Hoskin, Kenneth Lang, Charles A. Whitney, and Gerald Whitrow took part. A number of our referees have anonymously contributed important paragraphs to the text of this volume; they include Horace Babcock, James G. Baker, William McCrea, Howard Plotkin, and Charles A. Whitney. Among the other referees who have given invaluable assistance are Joseph Ashbrook, Bart J. Bok, David Dewhirst, Gérard de Vaucouleurs, David Edge, Eric Forbes, Dieter B. Herrmann, Richard Hirsh, Dorrit Hoffleit, Derek Howse, Henry King, Kevin Krisciunas, Nicholas Mayall, Donald Osterbrock, Robert Smith, John R. Shakeshaft, René Taton, Deborah J. Warner, Brian Warner, and Joan Warnow.

A variety of editorial assistance has been given by Barbara Welther, most significant of which has been assembling the illustrations and captions for the volume. Joan Jordan and Jennie Woodhouse have helped with typing or word processing. For all of these sources of aid the volume editor and the General Editor extend their warmest appreciation.

The authors of individual chapters or sections also wish to express their gratitude for specific help they have received:

John Lankford, Chapter 2, acknowledges grants from the Graduate Research Council of the Graduate School, University of Missouri-Columbia, which have partially funded his research; he also expresses his special thanks to Brenda Corbin (US Naval Observatory), Judith Lola (Yerkes Observatory), and J. Alexandre (Bibliothèque de l'Observatoire de Paris).

Albert Van Helden, Chapter 3, acknowledges the aid of John J. Zuger in the preparation of the chapter.

David DeVorkin, Chapter 4, expresses his appreciation to the archives and archivists who aided in this study including the American Institute of Physics Center for the History of Physics, the Princeton University Library and Archives, and the Harvard University Archives. Special thanks are due to A.J. Meadows, O. Gingerich, K. Hufbauer, S. Weart, C.A. Whitney and D. Woodward.

David Evans, Chapter 9, is particularly indebted to personal correspondence with Harley Weston Wood of Sydney, Jorge Sahade of Argentina, B.M. Lewis of Wellington, R.H. Stoy, now at Edinburgh, and G. de Vaucouleurs at Austin.

Woodruff Sullivan, Chapter 11, is grateful to the Program for History and Philosophy of Science of the US National Science Foundation for support of his research on the history of radio astronomy.

PART I

The birth of astrophysics and other late-nineteenth-century trends

(*c.* 1850–*c.* 1920)

1

The origins of astrophysics

A.J. MEADOWS

The rise of astrophysics in the nineteenth century represented a major turning point in the development of astronomy. In terms of the balance of present-day research, it proved to be the vital step forward. Prior to the growth of astrophysics, by far the major effort in astronomy was devoted to studies of where bodies were, and of how they moved. Now the emphasis has reversed: most of our current research effort is concerned with the physical properties of celestial bodies.

Astrophysics and spectroscopy

Physical studies of celestial objects can be carried out in a variety of ways; but the crucial factor in the early development of astrophysics was the study of spectra. The reason is simple – spectroscopy provided more information on physical and chemical properties than any other method. Interest in laboratory spectroscopy began to grow in the first half of the nineteenth century, as part of the general development of chemical techniques. Flame spectra provided a useful tool in some of the detailed studies of elements and simple compounds that characterized the rise of modern chemistry.

Astronomical spectroscopy can be seen, in part, as a later off-shoot of these initial attempts at chemical analysis via spectra. But the link was not entirely straightforward for it was mediated initially by the strong nineteenth-century interest in the nature of colour. An interplay between studies of colour, spectra and astronomy can be found in the work of several scientists in the first half of the nineteenth century. Fraunhofer is an obvious example. Less well-recognized, perhaps, are the similar interests of British scientists such as D. Brewster (1781–1868) and John Herschel (1792–1871) or, later in the century, of the Leipzig astronomer J.K.F. Zöllner (1834–82).

Brewster and Herschel both worked on all three topics for several decades. Brewster first became involved in these areas via an attempt to show that Newton's ideas on the prismatic spectrum were incorrect; whereas Herschel originally concerned himself with experiments on colour filters and their properties. These differing interests led both men to studies of flame spectra and also to the study of the solar spectrum – the latter, because the Sun provided much the brightest laboratory light source then available. At first, study of the Fraunhofer lines in the solar spectrum was regarded as ancillary to the use of sunlight as a source of different colours (an emphasis similar to that of Fraunhofer). But it was quite rapidly recognized that the spectral lines themselves were important, since they might provide a way of improving chemical analysis. From the 1820s onwards, this emphasis on analysis led to two types of investigation. The first attempted to understand the spectral changes that occured when different substances were subjected to a range of conditions in the laboratory. The second tried to reconcile observations of laboratory spectra with those of the solar spectrum.

It was assumed from early on in the study of laboratory spectra that the spectrum observed was in some sense characteristic of the substance producing it. The problem was discovering the exact link. Spectra of the same substance could soon be produced in more than one way – for example, by flame or by spark. Differences between the spectra produced by different methods suggested that even pure substances might not be characterized by a unique spectrum. This was also suggested by the apparently ubiquitous appearance of the Fraunhofer D line (subsequently shown to be double), which was observed regardless of the substance under examination. Part of this difficulty in interpreting spectra related to a lack of understanding of the theoretical basis, and we return to it later. Part stemmed from the fact that chemistry itself was still in a stage of rapid development. Thus, the widespread distribution of sodium – and so of the D lines – was shown in the later 1850s to be due to trace contamination by salt. This explanation resulted from the chemists' continuing attempts to isolate 'pure' substances. By about 1860 chemists were in general agreement that spectra could be used to characterize chemical substances uniquely, though detailed interpretation of spectral differences remained difficult.

The problem of comparing laboratory spectra with the solar spectrum hinged on the obvious difference that the former were in emission, whereas the latter was in absorption. Could the two types of spectrum be equated? Brewster's interest in coloured vapours led him quite early to an examination of the spectrum of nitrous oxide (a brownish-red gas). He found that sunlight, when passed through the gas, showed absorption bands. Brewster's explanation of his results – he thought the nitrous oxide spectrum explained the solar lines – was totally inadequate. But his experiment did point to the existence of a link between emission and absorption. The next step – identifying the same line in emission and absorption – was taken by J.B.L. Foucault (1819–68) in Paris in the middle of the nineteenth century. What he did was to test the supposed coincidence between the solar D line and the prominent yellow emission line observed in laboratory spectra:

As this double line recalled, by its form and situation, the line D of the solar spectrum, I wished to try if it corresponded to it, and in default of instruments for measuring the angles, I had recourse to a particular process.

I caused an image of the Sun, formed by a converging lens, to fall on the arc itself, which allowed me to observe at the same time the electric and the solar spectrum superposed; I convinced myself in this way that the double bright line of the arc coincides exactly with the double dark line of the solar spectrum.

But Foucault's experiment went beyond the establishment of coincidence: it also hinted at the way in which absorption and emission lines were interrelated. For he continues:

This process of investigation furnished me matter for

some unexpected observations. It proved to me, in the first instance, the extreme transparency of the arc, which occasions only a faint shadow in the solar light. It showed me that this arc placed in the path of a beam of solar light, absorbs the rays D, so that the above-mentioned line D of the solar light is considerably strengthened when the two spectra are exactly superposed. When, on the contrary, they jut out one beyond the other, the line D appears darker than usual in the solar light, and stands out bright in the electric spectrum which allows one easily to judge of their perfect coincidence. Thus the arc presents us with a medium which emits the rays D on its own account, and which at the same time absorbs them when they come from another quarter.

Foucault's observation, published in *L'Institut* on 7 February 1849, can be seen as the real beginning of a comparison between solar and laboratory spectra. Yet it was not properly followed up for another ten years. There was no single reason for this delay, but one problem is worth mentioning – the continuing element of doubt whether, or not, the Fraunhofer lines were formed on the Sun. In the early 1830s Brewster showed that some Fraunhofer lines darkened as the Sun approached the horizon, obviously suggesting that the terrestrial atmosphere was involved in their production. The remaining lines might be of solar origin, but, in the mid-1830s, even this was questioned. J.D. Forbes argued that absorption lines seen at the edge of the solar disc should be darker than the corresponding lines at the centre of the disc, because the slant height through the atmosphere would be greater towards the limb. Observations of the limb spectrum at an annular eclipse in 1836 showed no such difference, thus casting some doubt on a solar origin for the lines. Alternative explanations – for example, that the Fraunhofer lines were due to interference effects within the optical system – could not be entirely ruled out (though it is probably fair to say that the explanation in terms of interference phenomena is more interesting for its analogy with early criticisms of Galileo's telescopic observations than for its wide acceptance).

Despite such uncertainties, most scientists interested in Fraunhofer lines were not only prepared to believe that they were formed on the Sun, but, by the 1850s, were beginning to link the results of laboratory spectroscopy to the formation of the solar spectrum. Perhaps the most extended discussion during this decade was provided by A.J. Ångström, but the trend is well reflected in a short reminiscence by G.G. Stokes (1819–1903).

As well as I recollect what passed between Thomson [i.e. Lord Kelvin] and myself about the lines was something of this nature. I mentioned to him the repetition by Miller of Cambridge of Fraunhofer's observation of the coincidence of the dark line D of the solar spectrum with the bright line D of certain artificial flames, for example a spirit lamp with a salted wick. Miller had used such an extended spectrum that the 2 lines of D were seen widely apart, with 6 intermediate lines, and had made the observation with the greatest care, and had found the most perfect coincidence. Thomson remarked that such a coincidence could not be fortuitous, and asked me how I accounted for it. I used the mechanical illustration of vibrating strings which I recently published in the Phil[osophical] Mag[azine] in connection with Foucault's experiment. Knowing that the bright D line was specially characteristic of soda, and knowing too what an almost infinitesimal amount suffices to give the bright line, I always, I think, connected it with soda. I told Thomson I believed there was vapour of sodium in the Sun's atmosphere.[1]

Kirchhoff

Although speculations concerning the nature of the solar spectrum were common by the 1850s, one fundamental requirement for its proper discussion still remain unfulfilled. An explanation of the production of spectra required a basic physical understanding of the relationship between absorption and emission. This was first provided by G.R. Kirchhoff (1824–87), working with his Heidelberg colleague, R.W.E. Bunsen (1811–99). Bunsen's development of his well-known burner, with its non-luminous flame, was an important step forward in the method of producing spectra. However, Bunsen intended to use the flame in conjunction with colour filters: it was Kirchhoff who suggested the use of a prism. Between them, they managed to evolve the basic instrumental requirements for a proper spectroscopic analysis.

In the autumn of 1859 Kirchhoff and Bunsen effectively repeated and extended Foucault's work, though apparently neither was aware of

1.1. R.W.E. Bunsen, G.R. Kirchhoff and H.E. Roscoe in Heidelberg, 1862.

Foucault's experiments. (One of the problems of early astrophysics was that key results often appeared in publications not regularly scanned by others in the field.) In November 1859, Bunsen reported to the English chemist, H.E. Roscoe:

At the moment I am occupied by an investigation with Kirchhoff which does not allow us to sleep. Kirchhoff has made a totally unexpected discovery, inasmuch as he has found out the cause for the dark lines in the solar spectrum and can produce these lines artificially intensified both in the solar spectrum and in the continuous spectrum of a flame, their position being identical with that of Fraunhofer's lines. Hence the path is opened for the determination of the chemical composition of the Sun and the fixed stars with the same certainty that we can detect chloride of strontium, etc., by our ordinary reagents.[2]

Over the next two years, Kirchhoff established the basic principles of spectrum analysis. His results can be summarized as follows: (1) Incandescent solids or liquids typically give continuous spectra in the visible region, while gases give lines or bands (the positions of which depend on the gas examined). (2) When a source of continuous spectrum, such as the Sun, is viewed through a cool gas the wavelengths at which absorption occurs correspond to those at which emission occurs when the gas is heated.

These basic principles led Kirchhoff on to two immediate conclusions concerning the Sun. The first concerned its chemical composition, as derived from a comparison of solar and laboratory spectra.

Iron is remarkable on account of the number of the lines which it causes in the solar spectrum; magnesium is interesting because it produces the group of Fraunhofer's lines which are most readily seen in the Sun's spectrum, namely, the group in the green, consisting of three very intense lines to which Fraunhofer gave the name b. Less striking, but still quite distinctly visible, are the dark solar lines coincident with the bright lines of chromium and nickel. The occurrence of these substances in the Sun may therefore be regarded as certain. Many metals, however, appear to be absent; for although silver, copper, zinc, aluminium, cobalt and antimony possess very characteristic spectra, still these do not coincide with any (or at least with any distinct) dark lines of the solar spectrum.[3]

Kirchhoff's second deduction concerned the nature of the Sun. He postulated that the Sun consisted of an incandescent liquid core (he thought, at first, that it might have a solid crust) surrounded by an extensive gaseous envelope (which he identified with the corona). This model was soon replaced by astronomers with one that better fitted their observations. Nevertheless, it is important as the first model of the Sun based on the new physics (and chemistry) of the nineteenth century. The changes it introduced become evident from a comparison with the solar model previously accepted. Though not solely due to William Herschel, this model was often attributed to him in nineteenth-century writings. It supposed that the Sun possessed a cool, solid core surrounded by luminous cloud layers. Brewster, in his *More Worlds than One*, a popular work on life in the universe, indicated that this picture was still acceptable in the 1850s.

... we approach the question of the habitability of the Sun, with the certain knowledge that the Sun is not a red-hot globe, but that its nucleus is a solid opaque mass receiving very little light and heat from its luminous atmosphere.

After Kirchhoff, the assessment of what solar models were scientifically acceptable changed radically.

Solar eclipses

Kirchhoff's model emphasized the important role played by the solar atmosphere, thereby increasing interest in its properties. Since the atmosphere could only be observed at eclipses, the attention of astronomers was concentrated on these.

The main solar features visible at a total eclipse – the chromosphere, corona and prominences – were all noted before the 1860s. Interest in eclipse studies was stimulated by F. Baily's observations of bright 'beads' of light round the limb of the Moon at the 1836 eclipse. This led to more extensive observations at the next European eclipse in 1842, and consequent agreement on the existence of both prominences and the corona. The chromosphere was clearly distinguished at the 1851 eclipse. The physical nature of these features remained at first uncertain, but the advent of astronomical spectroscopy offered the possibility of a solution. A series of total eclipses at the end of the 1860s and in the early 1870s allowed repeated attacks on the problem using the newly-developed astrophysical techniques.

The first major advance occurred at the 1868 eclipse, which was watched from India and Malaya by several European astronomers. Spectroscopic observations of the prominences at this eclipse showed the presence of bright lines, so demonstrating that prominences were gaseous. P. J. C. Janssen (1824–1907), one of the observers, was sufficiently impressed by the brightness of the lines that he decided to scan the solar limb out of eclipse to see if the lines could still be seen. He was immediately successful; observations on succeeding days further showed that prominences could undergo rapid changes in size and shape with time. Meanwhile, others had already considered the possibility of observing prominences out of eclipse. Initially, the instrumentation available was insufficiently powerful, but in 1868 Norman Lockyer (1836–1920), using new equipment, managed to detect the spectra of prominences out of eclipse

1.2. The French mint struck this medal to commemorate the discovery by Jules Janssen and Norman Lockyer of the method of observing prominences out of eclipse.

almost simultaneously with Janssen. Since prominences were frequently thought of as jets from near the region of the solar surface, these spectroscopic results were taken as confirmation of the Sun's basically gaseous nature.

It soon became apparent that the red-coloured region seen round the limb of the Sun at eclipses also produced a bright-line spectrum that could be detected out of eclipse. Indeed, Father Angelo Secchi (1818–78) soon showed that hydrogen in emission could be traced all the way round the solar limb. As a result of these observations, the need to study prominences and the chromosphere at eclipses became much less pressing. Eclipse spectroscopy thenceforth concentrated on other problems, especially those relating to the corona and to the flash spectrum.

Because the corona was so faint, progress in its spectroscopic observation proved to be slow. The most important advance in the early years was the confirmation that it produced an emission-line spectrum. At the 1869 eclipse, visible in North America, the American astronomers C.A. Young (1834–1908) and W. Harkness independently confirmed the presence of a bright coronal line in the green region of the spectrum.

These advances in chromospheric and coronal spectroscopy raised a further query, this time concerning the absorption spectrum. Kirchhoff's

ideas implied that the part of the solar atmosphere producing the Fraunhofer spectrum should, when seen at the limb, show an identical spectrum in emission. Observations of the chromospheric and coronal spectra clearly demonstrated that neither of these was the required reversal of the Fraunhofer spectrum. Where, then, was it produced? This question finally received an answer at the 1870 eclipse. Young had looked for the reversed spectrum at the previous eclipse, but had failed to find it. This time, using a tangential rather than radial slit, he was successful. He observed the brief appearance of a set of emission lines – apparently at complementary wavelengths to the normal absorption spectrum. The short duration indicated that the part of the solar atmosphere producing the lines must be of small vertical extent. Young's observation therefore not only confirmed the accuracy of Kirchhoff's original prediction, but also pinned down the region of the solar atmosphere that produced the Fraunhofer lines. This 'reversing layer', as it was called, occupied the region just above the supposed photospheric cloud layer.

By the early 1870s the main areas of astrophysical attack on the problems presented by the solar atmosphere had been determined. The following decades were almost fully occupied in pursuing the insights, and sorting out the confusions, that arose during this initial period of eclipse observation. Two examples, relating to the chemical composition of the atmosphere, illustrate this point. Janssen and Lockyer had both noted a bright yellow line in prominence spectra near, though not identical with, the D line. It proved impossible to find an equivalent line in any laboratory source, and Lockyer boldly speculated that a new element – helium – was needed to explain it. From the beginning this explanation met with opposition – even from Lockyer's collaborator, E. Frankland. The debate continued until nearly the end of the century, when helium was finally detected in the laboratory. The green coronal line, which has been mentioned above, had an even more chequered history than the helium line. It was initially identified with a known iron line, an identification that caused increasing difficulty. Atmospheres were expected to show some segregation of their constituents, with the lightest molecules congregating to a greater extent near the top. The lightest known element was hydrogen, but the green line could certainly be traced to distances further from the Sun than any hydrogen line. How could the heavy element, iron, extend further into space than hydrogen? It was later shown that the line had been misidentified, but the basic question still remained – what element could it be? The solution of the problem was not found until the 1940s, when it was shown that this coronal line, and others, were produced by known, but highly ionized elements. Meanwhile, nineteenth-century astrophysicists attributed the lines to a new element – 'coronium' which, like helium, had not yet been detected on Earth.

Progress in the initial development of astrophysics

Early astronomical spectroscopy concentrated on the Sun, in part because it was the brightest object available. But another factor lay behind this special interest in the Sun – the discovery of solar–terrestrial relations. S.H. Schwabe in Germany had begun systematic observations of sunspots in 1826. By 1843, he was able to announce that the number of spots visible seemed to undergo a cyclical variation with period of about ten years. The conclusion attracted little attention until it was quoted by Humboldt in his book *Kosmos*. This appeared in 1851, at just the time that J. v. Lamont in Germany and E. Sabine in England were analys-

ing the results of several years' geomagnetic observations (the former for two German stations; the latter for four, widely distributed, colonial stations). Both found that geomagnetic variations occurred with a period of about ten years. It was a small step from this to the conclusion that some link existed between changes on the Sun and on the Earth.

Attention was thus drawn to the physical properties of the Sun, and more especially to the nature of solar activity. As John Herschel wrote to Faraday on 10 November 1852: "If all this be not premature we stand on the verge of a vast cosmical discovery such as nothing hitherto imagined can compare with." Many astronomers were consequently stimulated to observe sunspots in the 1850s and 1860s. During the latter decade, the interest in solar–terrestrial relations mixed with the new solar spectroscopy in the development of astrophysics.

An excellent example of this mixture of factors in the development of astrophysics is provided by Secchi's career. Initially attracted by Joseph Henry's photometric observations of sunspots in the 1840s, the Vatican astronomer moved on to a more general interest in the Sun during the 1850s and then to spectroscopic studies of the Sun and stars in the 1860s. The connection between the growth of his interest in solar physics and in terrestrial magnetism was noted with distaste in his obituary in *Proceedings of the Royal Society of Edinburgh*:

Secchi, though an excellent observer and a man of great power, was of a discursive turn of mind. He had little power of concentration, and appears to have tired of the monotony of astronomical observations, and to have turned his attention to the more popular studies of terrestrial magnetism and solar physics.

In view of this emphasis, it is hardly surprising that the first major attempt to provide a comprehensive explanation of the new solar observations – by H. Faye – started from recent studies of sunspots. Although the details of Faye's model soon came to be queried (and were abandoned to some extent even by their author), his general approach indicated the framework for most later discussions of solar physics. Two points, in particular, were widely accepted and were, in fact, also proposed by John Herschel and Secchi. First,

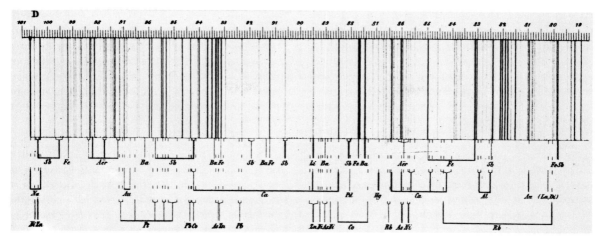

1.3. A section of the solar spectrum drawn by Kirchhoff and published in 1863 with the lines identified for such metals as antimony, iron, barium, lithium, sodium, gold, calcium, palladium, mercury, aluminium, lanthanum, and didymium (a mixture of neodymium and praseodymium, formerly thought to be an element).

the temperature increase from the surface inwards was sufficient to make the whole body of the Sun gaseous. Second, the photosphere represented a level in the Sun where refractory substances could condense out, forming a cloud layer. In establishing these propositions, a good deal of weight rested on the sunspot observations – visual, photographic and spectroscopic. For example, one important piece of evidence was the spectroscopic detection of increased absorption in sunspots by Lockyer and Secchi in the latter half of the 1860s.

It was an obvious extension of Kirchhoff's work on the total solar spectrum to examine the spectra of specific solar features, such as sunspots. Equally obviously, the sort of chemical and physical analysis that Kirchhoff had carried out on sunlight could also be applied to the light from stars and nebulae. One initial problem in all this work, which led to a number of false starts, was the difficulty of determining the precise wavelengths of lines in the spectra of celestial bodies.

Kirchhoff's original map of the solar spectrum suffered from a non-linear wavelength scale, owing to its prismatic dispersion. His arbitrary scale divisions were therefore not readily reproducible on instruments elsewhere. At the end of the 1860s Ångström published a new map of the solar spectrum, based on measurements with a grating, which gave a normal dispersion. His calibration, using a unit of one ten-millionth of a millimetre (one Ångström), soon displaced

Kirchhoff's scale for the intercomparison of spectra.

The main requirement for astronomical spectra was not simply that they should be accurately calibrated, but also that they should be of the highest dispersion possible. The basic aim behind many astronomical measurements was the wish to determine chemical composition. This involved a comparison of the positions of lines from laboratory and astronomical sources. The higher the dispersion, the more readily any lack of coincidence could be detected. Unfortunately, the higher the dispersion, the fainter the resultant spectrum. All early measurements were visual; even with an object as bright as the Sun, Kirchhoff strained his eyes in drawing the spectrum. (His map of the solar spectrum was completed by one of his pupils.) For stars, the problem was obviously increased. Of the two astronomers most involved in stellar spectroscopy in the 1860s – Secchi and William Huggins (1824–1910) – the latter was more concerned, since he used the higher dispersions. In his earliest paper on stellar spectra, in *Philosophical Transactions* in 1864, Huggins pointed out:

The investigation of the nature of the fixed stars by a prismatic analysis of the light which comes to us from them, however, is surrounded by no ordinary difficulties. The light of the bright stars, even when concentrated by an object-glass or speculum, is found to become feeble when subjected to the large amount of dispersion which is necessary to give certainty and value to the comparison of the dark lines of the stellar spectra with the bright lines of terrestrial matter. Another difficulty, greater because it is in its effect upon observation more injurious, and is altogether beyond the control of the experimentalist, presents itself in the ever-changing want of homogeneity of the Earth's atmosphere through which the stellar light has to pass. This source of difficulty presses very heavily upon observers who have to work in a climate so unfavourable in this respect as our own. On any but the finest nights the numerous and closely approximated fine lines of the stellar spectra are seen so fitfully that no observations of value can be made.

It was hoped from the start that both the problems noted by Huggins might be overcome by substituting the photographic plate for the human eye. Huggins attempted to photograph stellar spectra as early as 1863, but no results of value were obtained until the following decade. He was nevertheless able to establish two basic points about stellar spectra from visual observations. First, the same chemical elements could be detected in stars as were also found on the Sun and Earth. Second, the elements detected were not necessarily identical in different stars (for example, Huggins detected hydrogen in Aldebaran, but not Betelgeuse).

The difficulty of observing stellar spectra visually slowed down not only the chemical analysis of stars, but also the application of what was recognized to be a powerful new method for studying stellar motions. Christian Doppler had suggested in the 1840s that motion of a body in the line of sight could affect its colour. A few years later, A.-H.-L. Fizeau examined this possibility more rigorously, pointing out that the Fraunhofer lines might provide a delicate test of such motion. In 1868, both Huggins and Secchi tried to detect such a 'Doppler' shift in the spectrum of Sirius. Secchi's results were inconclusive, but Huggins, using better instrumentation, was able to claim he had found a definite red shift in the spectrum. Conclusive proof of Doppler effects in celestial spectra came, however, in 1871 from solar observations. H. C. Vogel, who was working at F. G. von Bülow's observatory near Kiel, then reported in *Astronomische Nachrichten*:

On 9 June, with the aid of a reversing spectroscope

made available by Professor F. Zöllner which had been fitted on the large equatorial telescope, we succeeded in demonstrating the rotation of the Sun through the displacement of the spectral lines... From the observations, the velocity would seem to be rather greater than has up to now been deduced from the movements of the sunspots. However, in some respects these observations themselves have a considerable degree of uncertainty, whilst in other respects the determination of the wavelength of individual lines in the solar spectrum is not yet so exact for observations of the kind under consideration that one would be justified in drawing any more far-reaching conclusions. Only the conclusion, which is not unimportant for hypotheses about the nature of light, that the motion of a luminous point can lead to a change of wavelength for the light beams emitted from it, can be regarded as certain.

Of all early spectroscopic results, perhaps the most frequently cited single observation is Huggins's examination of a planetary nebula in Draco. The resultant detection of its emission spectrum necessarily implied, from Kirchhoff's arguments, that the nebula must be gaseous. Huggins initially identified the brightest nebular line as being due to nitrogen; but subsequent observations showed that, whereas the nitrogen line was double, the nebular line was single. In the 1880s, Lockyer pressed the claim of magnesium, rather than nitrogen, as the source of this line (his identification being linked to the meteoritic hypothesis of stellar evolution which he was then developing). Huggins disputed this, and wrote to J. E. Keeler at Lick Observatory asking him to use the better facilities there to resolve the question. Keeler's observations in 1890 showed that no known laboratory line satisfactorily explained the properties of the chief nebular line. By the end of the nineteenth century, it was therefore supposed that, just as the corona contained 'coronium', so the nebulae contained 'nebulium'.

The discovery of the nebular emission lines is often seen as a vital turning point in the historical development of ideas concerning the physical nature of the nebulae. Although there is a good deal of truth in this, it must also be recognized that the application of spectroscopy to nebulae led to ambiguous results. Huggins's original observations showed that only a third of the nebulae observed had bright lines: the remainder possessed

1.4. William Huggins in his observatory at Tulse Hill, 1890s. His spectroscope employed a 4-inch grating from Brashear, and was attached to a 15-inch Grubb refractor lent by the Royal Society.

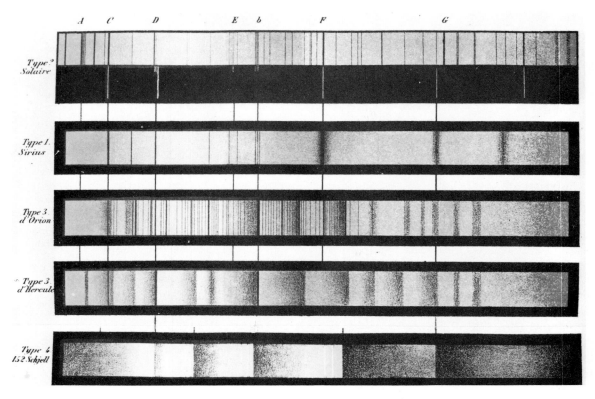

1.5. Secchi's spectral classes, reproduced from a pair of colour plates in his *Les Étoiles* (Paris, 1879). Type 2 is the Sun; Type 1 is Sirius; Type 3 (upper) is Betelgeuse; Type 3 (lower) is Alpha Herculis; Type 4 is 152 Schjellerup = Y CVn.

continuous spectra. Since his list was biased in favour of emission-line nebulae, subsequent investigations only served to confirm that such objects were in a minority. In consequence, the early spectroscopic observations of nebulae tended to complicate the accepted picture, rather than simplify it. In view of the faintness of these objects, knowledge of nebular spectra grew more slowly in the nineteenth century than any other branch of astronomical spectroscopy.

For the most part, ability to interpret the spectra of objects outside the solar system depended, in the early days of astrophysics, on the extent to which laboratory and solar observations could provide a parallel. For example, the first spectroscopic observations of a nova – in 1866 – showed both a gaseous and a stellar spectrum. These were compared with the emission and absorption spectra currently being observed on the Sun. The Fraunhofer spectrum was, of course, well known; what was discovered for the first time in the 1860s was the existence of solar emission lines.

Astrophysics and classification

One of the obvious characteristics of early astrophysics was its emphasis on classification. In general, such classification proved complex, because several different factors usually needed to be taken into account simultaneously. We can illustrate this by developments in the classification of stellar spectra.

Two methods of obtaining spectra were developed during the 1860s. The first, with the spectroscope attached to the eye-end of the telescope, was designed to provide high-dispersion spectra (in order, for example, to study the chemical composition of stars). In the second, the prism was placed in front of the telescope objective, producing a spectrum of lower dispersion, but correspondingly greater brightness. This proved useful for classification since more stars could be surveyed more rapidly. Secchi adopted this second method, using a dispersion just sufficient to distinguish the main features of a stellar spectrum, and recording what these were for all the brighter stars. He found that the vast majority of stellar spectra could be classified into three main categories (one of which he subdivided into two).

The main problem lay not in the classification, but in the identification of the physical factors that caused the spectral differences. Were they due to temperature, pressure, chemical composition, or to some combination of these? It was recognized from early on that some stars could not be fitted into this simple classification scheme, but with no firm idea of the factors at work, the reasons for such oddities remained a matter for speculation.

Classification is a typical preliminary stage of any science. Astrophysics was inevitably compared in this respect – to its disadvantage – with classical astronomy, which had long left the stage of classification to become a quantitative, interpretive science. Indeed, astrophysics fared badly in such a comparison not only with contemporary astronomy, or with the physical sciences as a whole, but even with biology, where Darwinian theory was being fiercely debated in the period when astrophysics was born. So, although astrophysics was recognized as a new and exciting subject, it was simultaneously seen as one with a less-than-secure theoretical base. One of the great problems of astrophysics in the nineteenth century was that it remained in this position of arrested growth. The observational data accumulated, but agreed interpretations of the data came more slowly. As a recent writer has put it, "Some indication of the growth rate of astrophysics can be derived from the number of research articles in the field. By the last decade of the nineteenth century this had only risen to about five per cent of the total publications in astronomy: the major expansion of astrophysical literature had to await the twentieth century."

One factor helped during this period: the area of astrophysics that could be put on a quantitative basis from the beginning – radial velocities – overlapped most in interest with classical astronomy. It was therefore the part of astrophysics that could be most readily absorbed into classical astronomy, and, to that extent, paved the way for further collaboration between the two branches.

Factors in the development of astrophysics

The growth of astrophysics in its early stages depended on the activities of a small number of scientists. What in their backgrounds and interests led them towards the study of astrophysics? Some were scientists with wide-ranging interests, one of which was classical astronomy. An obvious example is John Herschel, whose investigations covered many topics in astronomy, physics and chemistry. For such a person, the cross-fertilization of ideas that led to an interest in astrophysics is easy enough to understand. Others of the early pioneers had a relatively limited experience of astronomy prior to their entry into astrophysics, but a good knowledge of physical science – Kirchhoff is an excellent example. Yet another type of person, though committed to astronomy, was less interested in routine observation and calculation, and, in some cases, lacked the mathematical training of the traditional astronomer. Huggins, who was one such person, reflected in 1897:

I soon became a little dissatisfied with the routine character of ordinary astronomical work, and in a vague way sought about in my mind for the possibility of research upon the heavens in a new direction or by new methods. It was just at this time, when a vague longing after newer methods of observation for attacking many of the problems of the heavenly bodies filled my mind, that the news reached me of Kirchhoff's great discovery of the true nature and the chemical constitution of the Sun from his interpretation of the Fraunhofer lines.

This news was to me like the coming upon a spring of water in a dry and thirsty land. Here at last presented itself the very order of work for which in an indefinite way I was looking.

Yet the new science required its own kind of expertise – in spectroscopy – and some of the entrants lacked knowledge in this area, too. One result was that, during the 1860s, there were important instances of joint activity between scientists interested in the astronomical application of spectroscopy, on the one hand, and trained laboratory spectroscopists, on the other. Kirchhoff worked with Bunsen, Huggins teamed up with W. A. Miller, and Lockyer called on the assistance of Frankland. After the early stages, though interaction between laboratory and astronomical spectroscopy continued at a high level, astrophysics had developed sufficiently for the importation of outside experts to be less necessary.

As with the older astronomy, astrophysics from the start required the assistance of another group of experts – the instrument-makers. Progress in astronomical spectroscopy obviously depended on the production of spectroscopes which could be used and adjusted while attached to the end of a telescope. Some of the advances needed for this were not strictly confined to astronomy. For example, improvements in collimation resulted from developments in other fields. But astrophysics, with its emphasis on compactness and lightness, was important in such areas as the improvement of multi-prism spectroscopes. At the same time, the new instrumentation was relatively cheap.

It is true that the practice of astrophysics required more than a spectroscope attached to a telescope. Huggins, reminiscing about the 1860s, remarked:

Then it was that an astronomical observatory began, for the first time, to take on the appearance of a laboratory. Primary batteries, giving forth noxious gases, were arranged outside one of the windows; a large induction coil stood mounted on a stand on wheels so as to follow the positions of the eye-end of the telescope, together with a battery of Leyden jars; shelves with Bunsen burners, vacuum tubes, and bottles of chemicals, especially of specimens of pure metals, lined its walls.

But even this additional equipment did not place early astrophysics (and especially solar physics) beyond the purse of any moderately well-off amateur astronomer.

Astrophysics was seen by many classical astronomers as a subject both physically and conceptually messy. Even when they appreciated the advances it had made, they were suspicious of its effects on the traditional practice of astronomy. In 1864 Admiral W.H. Smyth, a distinguished amateur astronomer and a close friend of Huggins, commented in print:

With all my admiration of the marvellous and extensive power of Chemistry in disintegrating the nature and properties of the elements of matter, I really trust it will not be exerted among the Celestials to the disservice or detriment of measuring agency; and this I hope for the absolute maintenance of GEOMETRY, DYNAMICS and pure ASTRONOMY.

What this quotation particularly underlines is the close association then seen to exist between chemistry and astrophysics. As the name 'astrophysics' implies, astronomers ultimately agreed that physics is more relevant to their subject than chemistry. This was less obvious in the nineteenth century – when many scientists counted spectroscopy as a branch of chemistry – than it may be today (although the name 'astrophysics' was coined early on in the development of the subject by Zöllner). But there was a deeper reason for this rapport between nineteenth-century astrophysicists and chemists than their mutual interest in chemical analysis: they also tended to have similar views on the ultimate nature of matter.

Although physical and chemical concepts must obviously be compatible in the final analysis, there were clearly discernible 'physical' and 'chemical' views on atoms in the nineteenth century. 'Physical' atoms were fairly simple entities: possessing little internal differentiation. Chemical atoms, on the contrary, were visualized as more complex, since they were required to have properties that could distinguish the various elements from each other. The number and variety of spectral lines observed naturally suggested that the atoms producing them must have a complex internal structure. Astrophysicists therefore gravitated towards the chemists' picture of an atom. Unfortunately, a quantitative understanding of spectra required developments in physics, not in chemistry. Astrophysics could therefore only make rapid progress when physics itself had advanced sufficiently. If the advent of astrophysics was a revolution in astronomy, comparable in some sense with the advances of seventeenth-century astronomy, then we must see Kirchhoff as playing a role similar to Kepler's. The equivalent conceptual breakthrough to Newton's *Principia* was the arrival of the Bohr atom in the early twentieth century.

Notes

1. Quoted in H.E. Roscoe, *The Life and Experiences of Sir Henry Enfield Roscoe* (London, 1906), p. 71.

2. *Ibid.*, p. 81.

3. Letter from G.R. Kirchhoff [to O.L. Erdmann] on the chemical analysis of the solar atmosphere, published in translation in *Philosophical Magazine*, ser. 4, vol. 21 (1861), 185–8.

Further reading

Agnes M. Clerke, *A Popular History of Astronomy during the Nineteenth Century* (4th ed., London, 1902)

William Huggins, The new astronomy: a personal retrospect, *The Nineteenth Century*, vol. 41 (1897), 907–29

J. Norman Lockyer, *Contributions to Solar Physics* (London, 1874)

A.J. Meadows, *Early Solar Physics* (Oxford, 1970)

Henry E. Roscoe, *Spectrum Analysis* (4th ed., rev. by the author and A. Schuster, London, 1885)

H. Schellen, *Die Spectralanalyse* (2nd ed., Brunswick, 1871), transl. by Jane and Caroline Lassell as *Spectrum Analysis* (London, 1872)

2

The impact of photography on astronomy

JOHN LANKFORD

Two technological developments helped transform astronomy during the second half of the nineteenth century. The spectroscope provided information on the physical and chemical composition of the stars, while photographic plates replaced the human eye and, coupled with spectroscopes, formed the technological foundations for astrophysics. There was, however, nothing inevitable about the application of photography to astronomical research. Technical and social factors determined the course of events. Improvements in photographic materials and the solution of numerous problems connected with photographic instrumentation made the use of photography increasingly attractive. A handful of pioneering amateurs led the way, followed by some highly respected professionals whose task it was to convince the rank and file that the new research technology was actually a useful tool. This process of experimentation, demonstration and education within the astronomical community extended over five decades after 1840. Only towards the end of the 1880s did photography acquire the status of a legitimate research tool. The aim of this chapter is to explain the sequence of events leading to the acceptance of photography by the astronomical community and to discuss significant applications of the new research technology.

Early experiments with astronomical photography, 1840–60

The first epoch in the history of astronomical photography occupied the decades between March 1840 when the Anglo-American chemist J.W. Draper (1811–82) obtained the first successful daguerreotype of the Moon and July 1860 when the wealthy British amateur W. De la Rue (1815–89), in cooperation with the Italian professional Father Angelo Secchi (1818–78), used

photography to demonstrate the solar origins of the so-called protuberances (solar prominences) seen as totality. The 1840s were a period of 'firsts' – isolated successes in daguerreotyping the Moon, a solar eclipse (G.A. Majocchi in 1842), the solar spectrum (J.W. Draper in 1843) and the Sun (J.B.L. Foucault and A.-H.-L. Fizeau in 1845). At this early date a few imaginative astronomers like D.F.J. Arago (1786–1853), director of the Paris Observatory, envisaged a time when photography would be applied to selenography, photometry and spectroscopy. But to Arago's prophetic vision these applications of photography were only the beginning: "After all, when observers apply a new instrument to the study of nature, what they had hoped for is always but little compared with the succession of discoveries of which the instrument becomes the source – in such matters, it is on the unexpected that one can especially count." (*Oeuvres*, vol. 7 (1858), 500).

For a dozen years after it was made public in 1839, the daguerreotype remained the standard photographic material. A highly polished silver plate was fumed with iodine vapour. Researchers soon discovered that exposure to bromine vapour increased its sensitivity. After a second iodine vapour treatment the plate was ready for use. Following exposure, the plate was developed over a dish of mercury heated to 75 °C, and a bath of sodium thiosulphate was used to fix the image. Then, in 1851, F. Scott-Archer introduced the first wet collodion process. Essentially, the process consisted of coating a carefully prepared glass plate with 'collodion', a solution of gun-cotton and potassium iodide (or some other iodide) in alcohol and ether. It took considerable practice to achieve the manual dexterity necessary to coat a plate evenly. Once the collodion dried the plate was dipped into a silver nitrate solution, which converted the iodide to silver iodide. Plates had to be exposed while still wet, thus limiting exposures to ten or at most fifteen minutes. Yet Scott-Archer's process provided roughly a tenfold gain in sensitivity over the bromized daguerreotype.

Using the great 15-inch (38-cm) visual refractor at Harvard and with the assistance of commercial photographers, G.P. Bond (1825–65) made daguerreotypes of the Moon in 1849. Once he discovered that the visual and photographic foci differed, Bond was able to secure images that won a medal at the London Crystal Palace Exhibition of 1851. Equipped with wet plates and a new driving mechanism for his telescope, Bond resumed his photographic researches in 1857.

Bond's experiments came to an end in August 1860, but during the preceding three years his pioneering investigations paved the way for photographic astrometry and photometry. On the basis of sixty-two exposures of the double star Mizar, carefully examined in what may have been the first plate-measuring machine, he reported a high degree of consistency between distances measured on plates taken at different dates. He also showed that the separation was independent of the length of exposure. Bond quickly perceived the astrometric significance of photography. He wrote that well-defined star images could be bisected with a probable error of less than 0″.01. Further, photographic plates provided a permanent record that could be consulted at leisure and under conditions far less conducive to error than those that often obtained in a cold, dimly-lit dome. Bond realized that in a few short minutes the photographic plate could record groups and clusters of stars that a visual observer working with a micrometer would take months to measure.

Using Vega, Mizar and Alcor as his test objects, Bond undertook studies that would prove fundamental for stellar photometry. These plates were taken with exposures of from 1 to 120 seconds with an aperture of from 1 to 15 inches (2.5 to 38 cm). Writing in 1858 in the *Astronomische Nachrichten* Bond noted that "one remarkable property exhibited in the formation of the image is that a certain definite exposure, depending on the brightness of the star, is required before any trace of light action can be detected". According to Bond, image size was proportional to length of exposure. He represented these measurements of image di-

ameter, y, by the equation
$$Pt + Q = y^2$$
where t is exposure time and the constants P and Q are different for each star and each plate. Bond concluded that stars could be classified photographically by means of the value of P. This could be achieved in several ways, the simplest of which "depends upon the time required by unequal stars to form equal images". He also suggested that "the reciprocal of the area of the object-glass affording equal images in equal exposures, will also be an independent measure of photographic magnitudes".

During the 1850s various tyros dabbled with astronomical photography, but after Bond at Harvard the only sustained investigations were those carried out by De la Rue. The British amateur was inspired to undertake his first experiments after viewing Bond's lunar daguerreotypes at the Crystal Palace. He built a 13-inch (33-cm) reflector and in 1852 began photographing the Moon using wet plates, but, discouraged because he had no driving mechanism, soon discontinued the work. However, in 1857 De la Rue acquired a drive and turned again to lunar photography. His best images (2.8 cm in diameter) stood enlargement to 20 cm. Employing techniques of stereoscopic projection, De la Rue used his plates for studies of lunar libration and of surface features, but unfortunately, these projection techniques often produced spurious results. De la Rue devoted a large portion of his 1859 report to the British Association to a step-by-step exposition of the actual techniques for preparing, exposing and developing astronomical wet plates, thus providing a detailed practical guide for others who might wish to enter the field. Further, he discussed photographic instrumentation and argued the case for reflectors because of their ability to focus all colours at a single point.

In response to De la Rue's 1859 paper, H. Faye (1814–1902) of the École Polytechnique reported to the Paris Academy "On the state of astronomical photography in France". Unlike De la Rue in London or Bond at Harvard, Faye had little to say concerning concrete accomplishments; the French, it seems, were full of ideas but as yet did not have much to show in the way of actual results. Faye did, however, make a point that would be justified by future events. The French, he contended, sought to apply photography to astrometry while the English were more interested in using it as a tool in descriptive astronomy (such as lunar mapping or the study of sunspots).

Systematic observation of the Sun by means of photography was urged by the doyen of English astronomers, John Herschel, and in 1854 the Kew Observatory Committee of the British Association requested De la Rue to estimate the cost of appropriate instrumentation. The Royal Society agreed to fund the project and De la Rue devised a photographic instrument along the lines suggested by Herschel. The Kew photoheliograph had an objective of 3.5 inches (8.9 cm) and a focal length of 50 inches (127 cm), giving a focal ratio of $f/14$. In practice it was stopped down to about 2 inches (5 cm). The objective was figured to bring the visual (yellow) and photographic (blue) rays into focus at approximately the same point. An ordinary Huygenian eyepiece placed at the focus of the objective provided an enlarged solar image of approximately 10 cm. But the projection system introduced aberrations into the enlarged image, and focusing became a critical problem; this operation was carried out by adjusting the objective rather than the eyepiece.

Perhaps the most difficult problem encountered by De la Rue involved the construction of a very rapid shutter mechanism to prevent overexposure. The final design included a sliding metal plate with an adjustable aperture placed just in front of the eyepiece. When released, the metal plate was drawn rapidly across the field by a rubber band. In order to activate the shutter, the retaining thread was set on fire; later it would be cut with scissors.

A daily photographic record of sunspots was made at Kew from 1858 to 1872, but De la Rue, whose gifts as an experimenter had already led to a number of important papers in physics and chemistry, was not content merely to collect sunspot data. In 1859 he began planning an expedition to Spain to photograph the coming solar eclipse, though even with the Kew instrument success was extremely doubtful. De la Rue concluded that at totality the amount of light available was about equal to that of the Full Moon. Yet, with an exposure of one minute, he found himself unable to secure an image of the Moon with the Kew photoheliograph. Observing at Rivabellosa, Spain, De la Rue decided to make visual measurements in case photography was a failure. As he recalled in his 1862 Bakerian Lecture before the Royal

Society, once the assistant reported that photography was successful, De la Rue abandoned visual work, knowing that the photographic plates would provide more accurate data. Later, comparing his plates with those taken at Desierto de las Palmas in Spain, about 400 kilometres away, the British amateur was able to demonstrate conclusively that the prominences were of solar origin.

Approaches to photographic instrumentation, 1860–80

After De la Rue's path-breaking applications of photography to astronomical research, the development of new forms of photographic instrumentation became a central concern. More often than not, these activities were carried out by amateurs. By the early 1860s two Americans were deeply involved in improving photographic instrumentation. L. M. Rutherfurd (1816–92) gave up the practice of law in order to devote full time to the design, construction and use of photographic instruments, measuring engines and devices to rule gratings for spectroscopic work. His fellow New Yorker Henry Draper (1837–82) earned a medical degree at New York University and served on its faculty, but his research field was astronomy. Both men were financially independent and spared no expense in creating the best-equipped optical and physical laboratories in America. Their work was marked by painstaking attention to detail. Draper once remarked that in the course of figuring a large mirror he covered an average distance of ten miles each evening as he walked the treadmill that provided power for his grinding and polishing machines. Both men were members of the National Academy of Sciences – high honour for amateurs in a scientific community obsessed with achieving professional status.

Rutherfurd equipped his observatory with an 11-inch (28-cm) Fitz refractor and soon became fascinated by Bond's photographic work. After extensive experiments, he still found that his photographic results fell far short of perfection. Discussing his research in 1865 in the *American Journal of Science*, Rutherfurd attributed the difficulty "to the uncorrected condition of the [visual] objective which diffused the violet rays over a large space". He devoted considerable time to experiments with various lenses inserted in the tube between the objective and the plate in order to

2.1. The Warren De la Rue party observing the 1860 total solar eclipse at Rivabellosa, Spain, with the Kew photoheliograph. The individual standing just to the right of the telescope is holding a lighted taper used to burn the thread that will release the shutter mechanism. At the far right an assistant is emerging from the darkroom.

correct the telescope for photography, but he found off-axis images badly distorted. He also tried separating the components of a 4-inch (10-cm) Clark objective, hoping to find a point at which the photographic and visual foci united.

By 1863 Rutherfurd had examined the photographic potential of a variety of instruments and found them all wanting. He then decided to construct an objective corrected solely for photography. Such an objective could not be tested visually, so Rutherfurd devised a method using the

spectroscope. With his 11-inch photographic refractor, Rutherfurd was able to record ninth-magnitude stars on wet plates in three minutes. He quickly realized the harmful relationship between his urban location and photographic seeing, commenting that only rarely in that situation did the plate capture all the details visible to the eye. Given the limitations of photographic materials then available, Rutherfurd was never able to do more than produce a permanent record of the stars easily visible with modest apertures.

Astronomers and historians have sometimes expressed surprise that Rutherfurd's stellar photographs received little recognition, while his lunar plates attracted a great deal of attention. The reason is clear. In the hands of a gifted engineer like Rutherfurd, photographic instrumentation developed far beyond the needs of most astronomers. Until the research programmes of professionals caught up with these innovations, pictorial representations of the lunar surface, suggesting the (largely illusory) possibility of quantitative analysis, received the lion's share of attention. Rutherfurd must have recognized this situation, for he worked to develop plate-measuring engines and to apply photography to astrometric studies. His photographs of clusters and star fields were measured with machines of his own design and construction. Late in life Rutherfurd turned over his instruments, plates and notebooks to Columbia University and in time most of the material was published.

Rutherfurd's interest did not stop with astrometry; he was fascinated by the new field of astrophysics and moved quickly from visual to photographic spectroscopy. Seeking maximum dispersion, he experimented during the 1860s with glass prisms and also with hollow prisms filled with foul-smelling carbon disulphide. Rutherfurd produced a series of fifteen overlapping plates that he assembled into a photographic map of the solar spectrum some two metres in length. Still not satisfied, the inveterate experimenter designed and constructed an engine to rule diffraction gratings on speculum metal; at its most successful, the machine produced gratings with 6809 lines per centimetre. Working with gratings during the 1870s, Rutherfurd prepared a second photographic map of the solar spectrum. This one involved twenty-eight plates of varying exposures and covered the region from below E in the green into the ultraviolet. Fully assembled, this photographic map of the solar spectrum was about three metres in length. These researches won lavish praise from professionals.

In 1864 the Smithsonian Institution published Draper's monograph, *On the Construction of a Silvered Glass Telescope, Fifteen and a Half Inches in Aperture, and Its Use in Celestial Photography*, which immediately took its place in the literature and remained in print for more than half a century.

Like De la Rue's 1859 report to the British Association, Draper's publication represented the fruits of exhaustive experimentation. Over a thirty-six-month period he devised a series of seven different grinding and polishing machines and produced more than one hundred mirrors ranging from less than $\frac{1}{2}$ inch to 19 inches (1 to 48 cm) in diameter. Had not Smithsonian Secretary Joseph Henry repeatedly urged Draper to commit his findings to paper, the young perfectionist might have postponed publication indefinitely, as he sought to contrive the perfect photographic telescope.

For his 15-inch (38-cm) mirror, Draper designed an elaborate altazimuth mounting that kept the eyepiece and plateholder stationary no matter what the altitude of the object under observation. Because of the form of mounting, he was forced to develop a system for moving the photographic plate to compensate for the Moon's apparent motion, a device that stands as the prototype of the double-slide plateholder so important for later photographic research with large reflectors. Solar photography was also part of his research programme. Convinced of the value of large instruments, he suggested that 48-inch (122-cm) or even 60-inch (152-cm) silver-on-glass reflectors were a possibility.

Draper devoted a great deal of effort to testing and comparing the photographic capabilities of reflectors and refractors. He constructed a 28-inch (71-cm) reflector in 1872 and three years later purchased a 12-inch (30-cm) Clark refractor, which was mounted as a counter-weight to the reflector. In 1880 this was exchanged for an 11-inch (28-cm) photographic refractor. Much of Draper's time was devoted to improving the mounting and driving mechanism. He built seven different clocks before he found one that met his demanding standards: the star must remain on the slit of the spectroscope for at least an hour when near the meridian.

Nowhere was Draper's passion for experimenting with various forms of photographic instrumentation more evident than in his approach to spectroscopy. His observing books show that between his first wet plate of the spectrum of Vega in May 1872, when he used the 28-inch reflector equipped with a quartz prism inside the Cassegrain focus, and the last dry plates exposed in August 1882 with the 11-inch photographic refractor

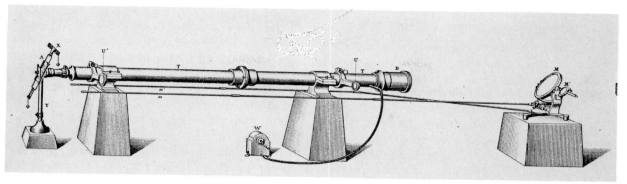

2.2. During the 1860s Laussedat used a form of this fixed long-focus photographic refractor to photograph solar eclipses.

fitted with quartz collimating lenses and an Iceland spar prism, Draper devised at least a dozen different instruments. They ranged from the slitless quartz prism of 1872, with which he soon obtained plates showing absorption lines, to a large spectrograph which employed a train of either three, six or nine prisms.

The instrument Draper used most frequently during the last years of his life included a Browning spectroscope with two 60° dense white flint prisms. The device was carefully braced, and an eyepiece permitted visual inspection in the yellow and red regions of the spectrum as a means of guiding. Plates, cut from commercial stock, were about 2.5 cm square, while the spectrogram measured about 1.58 mm wide and 12.7 mm long, extending from a point in the blue between G and F into the ultraviolet. Exposures ranged from 5 seconds to 228 minutes. In 1882 Draper was able to record the spectra of tenth-magnitude stars in the region of M 42. He also secured plates of cometary spectra.

At his death, Mrs Draper gave a collection of seventy-eight spectrograms to professional astronomers for measurement and discussion. Curiously, despite the years he devoted to building large reflectors, Draper apparently found refractors more suitable for photographic spectroscopy, for nearly two-thirds of the plates were exposed in either the 11-inch or 12-inch Clark instruments. Later his widow endowed the Henry Draper Memorial at Harvard College Observatory, which led to the great spectrographic survey culminating in the *Henry Draper Catalogue* (1918–24).

In Europe a young officer in the French Army Corps of Engineers, A. Laussedat (1819–1907), also made significant contributions to photographic instrumentation. Educated at the École Polytechnique, Laussedat early on sought to apply photography to topographic mapping. Laussedat taught at the École and later served as director of the Conservatoire des Arts et Métiers. The instrumentation he devised to photograph solar eclipses during the 1860s had a significant impact on French plans for observing the 1874 transit of Venus, and may also have influenced the American Transit of Venus Commission. He was elected to the French Academy in 1894. Between them, Laussedat, Rutherfurd and Draper were responsible for the most significant developments in photographic instrumentation in astronomy during the 1860s.

Laussedat's work reached the international astronomical community in March 1870 when Faye presented a memoir to the Academy on the application of photography to the coming transits of Venus. Faye quoted extensively from a letter in which Laussedat described his experiments, on the basis of which the young officer proposed a photographic refractor mounted horizontally in the meridian with a building to house the plateholder and provide a darkroom. Sunlight would be fed to the refractor by a siderostat. A transit instrument was to be placed behind the siderostat with a collimator to the rear. All rested on masonry piers and the transit was enclosed. The alignment of the siderostat and photographic instrument in the meridian was thus assured and deviations could be detected by means of the transit. Faye suggested that special photographic objectives of from sixteen to twenty metres' focus would be desirable because of the need for large images without recourse to eyepiece projection.

Photography and the transits of Venus

The transits of Venus in 1874 and 1882 offered astronomers the opportunity to organize an international campaign to improve the value of the solar parallax. In Britain and Germany there was considerable debate over the use of photography, but at length the Astronomer Royal and the German Transit of Venus Commission reluctantly agreed to include photographic observations as part of the 1874 programme. The French and American commissions were more receptive to the new research technology.

Because photography was so poorly understood and because there were so few precedents concerning its application to astronomical research, the international astronomical community could not reach consensus on instrumentation or photographic materials. British and German astronomers employed forms of De la Rue's photoheliograph, but expressed doubts concerning distortion which might be introduced by eyepiece projection. The American and French commissions opted for the long-focus refractor mounted horizontally. The American instrument was probably the most elaborate. Commission Secretary Simon Newcomb (1835–1909) of the US Naval Observatory spent considerable time in perfecting a design that would keep instrumental errors to a minimum. Especially important was the determination of plate scale. American plans called for doing this on the basis of exact measurements of the focal length of the objective using a fixed rod and special micrometer. Other expeditions sought to establish plate scale by measuring the photographic diameters of Venus and the Sun. British and German photographic parties used forms of dry collodion, while the French opted for daguerreotypes. The Americans employed wet collodion. All the major national expeditions engaged professional photographers to expose the plates, while the astronomers concentrated on visual observations.

Measurement, analysis and discussion of the 1874 photographs proved virtually impossible. In 1878 the Astronomer Royal admitted that the British photographic observations had been a failure. Because the solar limb was so indistinct, fading away gradually when magnified even a few diameters, no two assistants could agree on the same measurement; thus plate scale became impossible to determine. The American and French commissions published quantitative data derived from the plates, but declined to solve the equations of condition and derive a value for the solar parallax. The Germans became discouraged after attempts to measure the 1874 plates and published no photographic results at all.

The apparent failure of photography in the 1874 transit of Venus had profound consequences. In October 1881 scientists from fourteen nations gathered at Paris to plan for the 1882 transit. The president of the French Transit of Venus Commission was in the chair and a senior member of the German commission served as vice-president. After formal debate, the Congress recommended against photography for observing the coming transit. Thus the international astronomical community rejected photography as a research tool. In the main, professionals preferred traditional methods.

The dry plate revolution

As long as astronomers had to depend on wet plates they were limited in the uses to which they could put the new research technology. As early as 1854 French and British investigators were involved in experiments with dry plates, and by the mid-1860s significant progress towards dry collodion emulsion had been made. The 1874 transit of Venus was photographed with dry collodion by British and German parties. While there were several forms, the typical collodion dry plate involved the following. To a standard collodion mixture zinc bromide and a small quantity of nitric acid were added. The next step involved the addition of a solution of silver nitrate in water and alcohol. The result was an emulsion of silver bromide formed in suspension; this was allowed to ripen and then poured into a dish so that the solvents could evaporate. The result was a gelatinous substance that was washed to remove soluble salts. Once dried, the material was dissolved in alcohol and ether and applied to glass plates. The last step involved coating the emulsion with a preservative such as the beer–albumen mixture devised by W. de W. Abney (1843–1920), whose photographic experiments were of special importance to astronomers. Researchers quickly discovered that dry plates demanded new developers, and this line of enquiry expanded rapidly after

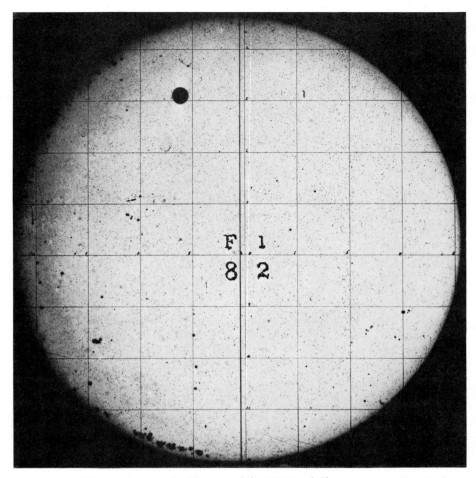

2.3. The 1874 transit of Venus photographed by one of the US Naval Observatory parties, station unknown.

about 1860. The introduction of the dry collodion plate led to a study of halation and by 1874 techniques for backing plates were in use. The sensitivity of these plates varied greatly.

So successful was dry collodion that researchers began looking for collodion substitutes. The result was the gelatin emulsion process in which silver salt is emulsified in gelatin. It was soon found that the speed of this new photographic material could be improved by several orders of magnitude either by prolonged exposure to heat or by the addition of ammonia. Thus, as the 1870s came to an end, new and exciting possibilities opened for astronomical photography through the use of rapid dry plates that could be exposed for as long as the astronomer cared to guide the telescope.

Even though the exposure times refer to terrestrial subjects, Table 2.1 illustrates improvement in photographic materials during the forty years after Arago announced the discovery of the daguerreotype.

The watershed of the 1880s

The 1880s stand as a watershed in the history of astronomical photography, for as the decade opened it was by no means certain that photography would play more than a peripheral role in astronomy. The acceptance of photography as a research technology during the 1880s was caused by several factors. Improvements in the sensitivity and quality of photographic materials must be assigned considerable weight, as must the development of new forms of specialized photographic instrumentation. But demographic factors were also significant. In the 1880s a new type of astronomer appeared: individuals, often trained in

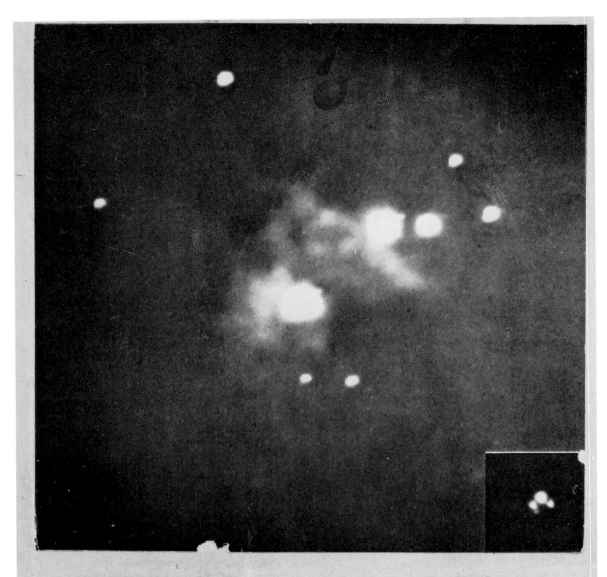

FIRST PHOTOGRAPH OF THE NEBULA IN ORION.

TAKEN BY PROFESSOR HENRY DRAPER, M.D.
September 30th, 1880, Exposure 51 minutes.
The picture is an Artotype enlargement by Harroun & Bierstadt from the original
negative. The large stars, being much brighter than the Nebula, are greatly over-
exposed. In the lower right hand corner is a photograph of the Trapezium alone,
with only 5 minutes exposure.

2.4. Draper's 1880 photograph of the Orion Nebula.

Table 2.1. *Comparative increase in sensitivity of photographic materials, 1839–80.*

Material	Minimum length of exposure for terrestrial subjects
Original daguerreotype	30 minutes
Collodion (wet plate)	10 seconds
Collodion emulsion (dry collodion)	15 seconds
Rapid gelatin emulsion (dry plates)	1/15 second

Source: Adapted from the article on photography, *Encyclopaedia Britannica* (11th ed., Cambridge, 1911), vol. 21, p. 491.

physics, whose concerns were not astrometric but rather with the fledgling specialty of astrophysics. These workers in the 'new astronomy' studied the physical conditions of the Sun and stars. To them, photography became an indispensable ally. Further, some students of the 'old astronomy' realized that astrometric research could be done efficiently by photography with no sacrifice of precision. Finally, in the 1880s photography dramatically demonstrated that the sensitive plate could 'see' more than the human eye. Astronomical photography moved beyond providing a permanent record of the sky at a given epoch to become a tool of discovery.

Early in the decade, however, amateurs continued to lead the way. In England the engineer A. A. Common (1841–1903) entered the field, and in 1883 secured plates of the Orion Nebula using a 36-inch (91-cm) reflector of his own design. Common's photographs showed marked improvement over those made by Draper only a few years before. On Draper's plates, M 42 was similar to the nebula seen visually, a bright central condensation together with outlying traces of nebulosity, whereas Common's photographs suggested a much more complex and extended object. In 1884, he won the Royal Astronomical Society Gold Medal for his photographic efforts. Later in the decade another British engineer joined the ranks of photographers using reflectors. Isaac Roberts (1829–1904) retired to Crowborough Hill, Sussex, and erected a private observatory equipped with an f/5 20-inch (51-cm) reflector. For twenty years

he produced photographs of nebulae and star clusters that were universally admired and that fully illustrated the capabilities of photography for the representation of nebulae. Both Roberts and Common were elected to the Royal Society.

Photography first attracted the serious attention of professional astronomers in a curious and roundabout fashion. David Gill (1843–1914), director of the Royal Observatory at the Cape of Good Hope, made a series of carefully guided plates of comet 1882b, working with a commercial photographer and a camera equipped with a 2.5-inch (6.3-cm) Dallmeyer portrait lens (f/4). While the object of these experiments was, of course, the comet, both Gill and European astronomers who examined his prints were struck by the number and quality of star images in exposures lasting up to an hour and forty minutes. Had it not been for Gill the matter might have rested here – another set of pretty pictures without apparent scientific value. Although Gill devoted his professional career to astrometry, he was sympathetic to new research technologies and fields of investigation. Indeed, his earliest work as an amateur in Scotland in the 1860s had involved lunar photography. By the 1880s Gill had become one of the dominant figures in Victorian science. As the astrophysicist A. S. Eddington recalled in 1915, "In some indefinable way he could inspire others with his enthusiasm and determination". After 1882 Gill brought all the force of his personality to bear in arguing the case for photography with other professionals.

In 1885 Gill received a grant from the Royal Society for research in astronomical photography. His primary goal was a photographic star map of the southern sky. In this mammoth undertaking he was soon joined by J. C. Kapteyn (1851–1922) of the University of Groningen, in the Netherlands, who offered to establish an astrographic laboratory and to measure, reduce and discuss plates exposed at the Cape. Both Gill and Kapteyn carried out these researches in spite of rigid financial constraints, for in 1887 the Royal Society terminated Gill's funding and he had to pay for the work out of his own pocket. The action of the Royal Society was significant: it was the result of opposition to astronomical photography among many professionals. As J. C. Adams, the Cambridge celestial mechanician, informed Gill, many of the

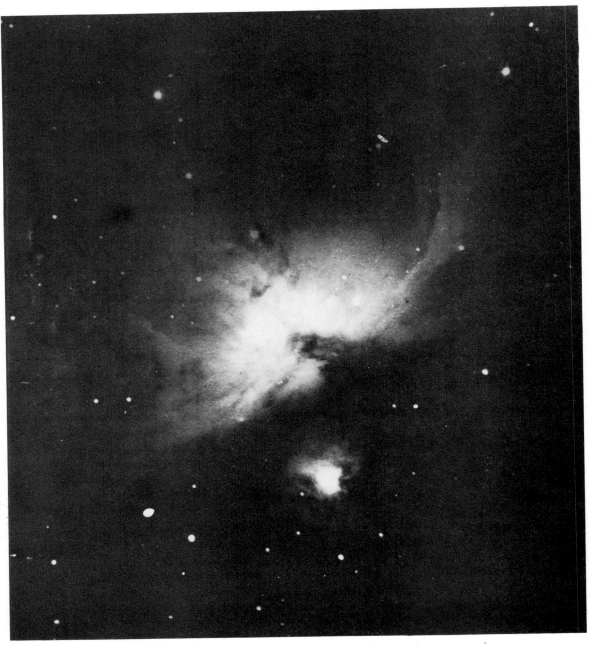

2.5. Photograph of the Orion Nebula taken by A. A. Common on 28 February 1883 with exposure of one hour.

practitioners of the old astronomy feared that photography would supersede meridian instruments. During the 1880s and 1890s the clash between the old and new astronomies frequently manifested itself in disputes over the scientific value of photography.

The *Cape Photographic Durchmusterung* (CPD) was executed using a 6-inch (15-cm) Dallmeyer objective of 4.6-ft (1.4-m) focal length (f/9). Gill's plan divided the sky into 612 fields. He employed two assistants, who carried out the photographic work between April 1885 and December 1890.

Each plate covered a field of 5° by 5°. Two separate exposures were made on different nights so that imperfections in the plates would not be confused with stars. The goal was to record all stars to visual magnitude 9.5. Plates improved so rapidly during these years that by 1887 Gill's assistants were able to reduce exposure times from an hour to thirty minutes. Using an ingenious measuring engine constructed from parts of several old telescopes, Kapteyn and his assistants measured plates by working a three or four hour shift in the morning and again in the afternoon. Each plate was measured twice and any discrepancies were then checked by Kapteyn himself. "For good and rich plates we generally made 300 to 400, occasionally even as many as 450 observations an hour", the Dutch astronomer wrote in the introduction to the *CPD*. The published catalogue contained the positions and photographic magnitudes of 454 875 stars. Kapteyn and Gill were among the first since Bond to engage in sustained research on such fundamental problems as the determination of photographic magnitudes and the measurement of stellar coordinates on photographic plates, and the *CPD* paved the way for the international *Carte du Ciel* project.

The early 1880s witnessed the completion of several major projects carried out by visual methods. E.S. Holden (1846–1914) at the US Naval Observatory published a monograph reviewing studies of the Orion Nebula. The next year H.C. Vogel (1841–1907) of the Potsdam Astrophysical Observatory issued a catalogue of the spectra of 4260 stars. E.C. Pickering's photometric catalogue of over 4000 stars appeared in 1884 from the Harvard College Observatory, and 1885 saw the publication of C. Pritchard's *Uranometria Nova Oxoniensis*. These elaborate visual studies, involving thousands of hours at the telescope, proved far from satisfactory. Holden could not demonstrate conclusively that changes had occurred in M 42. Vogel's research did not reveal stellar distribution patterns by spectral type as he had hoped. Pickering and Pritchard, using different instrumentation, found their photometric observations at variance in small but significant ways. The inconclusive results and discordant data produced by these great undertakings may have prompted astronomers to seek new research methods that would yield greater precision and provide a larger number of observations than traditional visual techniques. Holden, Vogel, Pickering and Pritchard all turned to photography during the second half of the decade, no doubt encouraged by the work of amateurs like Common and Roberts and professionals like Gill at the Cape or the Henry brothers at Paris.

Paul Henry (1848–1905) and Prosper Henry (1849–1903) joined the Meteorological Section of the Paris Observatory in 1864. Four years later they began construction of a 30-cm reflector as a hobby, and with this instrument undertook a visual map of the ecliptic. In 1871 the Observatory director, C.-E. Delaunay, arranged for the brothers to work full time on the ecliptic map. By 1884 the project was about one quarter complete, covering a zone 5° wide to the thirteenth visual magnitude. But the Henrys had come to the intersection of the ecliptic and the Milky Way and the stars became too numerous to map.

The Henrys turned to photography and quickly brought their mechanical and optical talents to bear. A 16-cm photographic objective was constructed and tested. In 1885 this was replaced by a 34-cm, f/10 photographic refractor. With this instrument the Henrys discovered the Maia Nebula in the Pleiades; only later was the object confirmed visually with the Pulkovo 30-inch (76-cm) refractor. Using long exposures, they recorded up to 2326 stars in the Pleiades region. These were amazing developments: an important astronomical object had been discovered by photography, and it was located in a region of the sky intensely scrutinized by generations of visual astronomers. In their experiments the Henrys enjoyed the support of Admiral E.B. Mouchez (1821–92), successor to Delaunay as director of the Paris Observatory. Mouchez was one of the first astronomers to appreciate the implications of Gill's 1882 photographs and to call them to the attention of others.

Impressed by the achievements of the Henrys, Mouchez began thinking about the possibility of a great photographic star chart. Writing in 1885 to Pickering, he suggested that such a map could be produced by five or six observatories over a period of six to eight years. In 1886 Gill wrote to Mouchez suggesting a plan of international cooperation that would employ photography to produce a catalogue of precise star positions as well as a photographic map covering the whole sky; apparently Otto W.

2.6. Paul and Prosper Henry with their astrographic camera in the Paris Observatory, mid-1880s.

Struve, director of the Pulkovo Observatory, made similar proposals. Gill urged that an international congress be called. Mouchez needed no further encouragement, and soon the French Academy issued invitations for an Astrographic Congress to convene in April 1887.

The Astrographic Congress and the *Carte du Ciel*

Fifty-six scientists from nineteen nations responded to the invitation. The very fact that the French Academy of Sciences called an Astrographic Congress served to lift the censure of astronomical photography voted by the 1881 Transit of Venus Congress. In the words of the official historian of the *Carte du Ciel*, "This development marked the systematic introduction of photography into astronomy". During eleven crowded days, the delegates agreed on a plan of international cooperation to produce a photographic map of the sky (to the fourteenth magnitude) and a precision catalogue for stars to the eleventh magnitude. After debate, the photographic refractor developed by the Henry brothers was adopted as the standard instrument. The specifications called for a 34-cm objective with a focal length of 3.43 m (f/10) so that one millimetre on the photographic plate represented one arc minute on the sky. The 16 × 16-cm plates covered a field 2° square. The optics were to be corrected with reference to the Fraunhofer G line. Guiding was accomplished with a 24.5-cm refractor of the same focal length as the photographic telescope and the steel tubes of both instruments were bound together to minimize flexure. Each photographic plate was to carry the imprint of a réseau, a network of fine lines at 5-mm intervals, which was to serve as a check against deformation of the photographic material and which later became indispensable in the measuring process. The image of the réseau was obtained by photographing a silvered-glass plate on which the lines had been carefully inscribed.

Cooperating observatories were to make two series of plates. Each plate destined for catalogue measurements would involve three exposures, of 20 seconds, 3 minutes and 6 minutes, and the telescope was to be moved after each exposure. Map plates were to be given longer exposures. Multiple exposures were designed to identify false stars caused by dust or defects in the plates. Both the catalogue and the map plates were to be executed in duplicate.

The 1887 Congress sought to establish uniformity in all areas of the work. Many questions, however, could not be decided without intensive research. Further, the project needed careful supervision, and so a Permanent International Committee was established by the Congress. This group was charged with the direction of the work and empowered to appoint special commissions and to resolve various problems as they arose. The Permanent International Committee met five times (1889, 1891, 1896, 1900, 1909) before the *Carte du Ciel* became Commission 23 of the International Astronomical Union (IAU) in 1919. After each meeting the French Academy of Sciences underwrote the publication of the *procès-verbaux des séances*. These five volumes also contained lengthy technical appendices. In addition, the Committee issued a *Bulletin* in seven volumes that appeared at irregular intervals between 1892 and 1917.

The Permanent International Committee of the *Carte du Ciel* project became the major forum for discussions of photographic astrometry and, to a lesser degree, of photographic photometry. Its work was reported, summarized and reviewed in all the European and American astronomical journals. The educational functions of the Committee would be difficult to overestimate: in the years from 1890 to World War I, it occupied a unique position in the international astronomical community.

Table 2.2 indicates the scope and division of labour of the *Carte du Ciel* project and the associated *Astrographic Catalogue*. Only in 1964 was the publication of the catalogue complete. Some observatories dropped out and their work had to be reassigned or shared, while others, like Potsdam, produced substandard results; some never attempted to secure funding for the map. As H.H. Turner, the Oxford astronomer who became first president of IAU Commission 23, suggested, the expense of engraving and printing the maps would cost each participating observatory between £5000 and £10000. Indeed, the project turned out to be more costly and time-consuming than anyone had foreseen. For example, Turner estimated the total cost of the Oxford portion of the astrographic catalogue at about £34000, includ-

Table 2.2. *Division of labour and execution of the* Carte du Ciel *project: catalogue and map.*

Zone centred on	Number of plates	Original observatory	Remarks	Map
+90° to +65°	1149	Greenwich	Catalogue completed	Published
+64° to +55°	1040	Vatican	Catalogue completed	Published
+54° to +47°	1008	Catania	Catalogue completed	Plates not taken
+46° to +40°	1008	Helsinki	Catalogue completed	Plates taken but not published
+39° to +32°	1232	Potsdam	Zones hastily published and then abandoned. Hyderabad, Uccle, Paris and Hamburg cooperated to reobserve, measure and publish this zone	Plates taken, printed and distributed by Uccle
+31° to +25°	1180	Oxford	Catalogue completed	Plates not taken
+24° to +18°	1260	Paris	Catalogue completed	Published
+17° to +11°	1260	Bordeaux	Catalogue completed	Published
+10° to +5°	1080	Toulouse	Catalogue completed	Published
+4° to −2°	1260	Algiers	Catalogue completed	Published
−3° to −9°	1260	San Fernando	Catalogue completed	Published
−10° to −16°	1260	Tacubaya	Catalogue completed	Published
−17° to −23°	1260	Santiago	Santiago withdrew and was replaced by Hyderabad, which completed the catalogue	Plates not taken
−24° to −31°	1360	La Plata	La Plata withdrew and was replaced by Cordoba, which completed the catalogue	Plates taken, but only zone −25° published
−32° to −40°	1376	Rio de Janeiro	Rio de Janeiro withdrew and was replaced by Perth, Edinburgh and Paris, which cooperated to complete the catalogue	Plates not taken
−41° to −51°	1512	Cape of Good Hope	Catalogue completed	Plates taken but not published
−52° to −64°	1400	Sydney	Catalogue completed	Plates taken but not published
−65° to −90°	1149	Melbourne	Melbourne withdrew after the catalogue plates had been taken and several zones published. Sydney and Paris cooperated to complete the catalogue	Plates taken but not published

Note: Total number of plates = 22054. (Had each participating observatory made both catalogue and map plates in duplicate, the figure would be 88216.)
Source: P. Couderc, "Historique de la Commission de la Carte du Ciel", *Trans. IAU*, vol. 14B (1971), 176–7.

ing the value of the astrographic refractor given by De la Rue and support from both the university and the government. Work on the Oxford zone (stars between +25° and +31°) was carried out between 1892 and 1903. Lacking the full state funding accorded to Greenwich or Paris, Oxford measured the rectilinear coordinates for only the 6-minute exposures, reversing the plate in the measuring engine between the first and second readings. Further, in star-rich portions of the zone, Oxford did not attempt to measure all stars of the eleventh magnitude.

While American astronomers attended the 1887 Congress and served on various commissions appointed by the Permanent International Committee, no observatory in the United States constructed an astrographic refractor and participated in the project. Even the US Naval Observatory, whose interests were primarily astrometric, made little effort to become involved beyond sending an observer to Paris in 1887. Part of the explanation of American lack of interest can be found in the comparatively low level of government funding for the Naval Observatory. But there were other

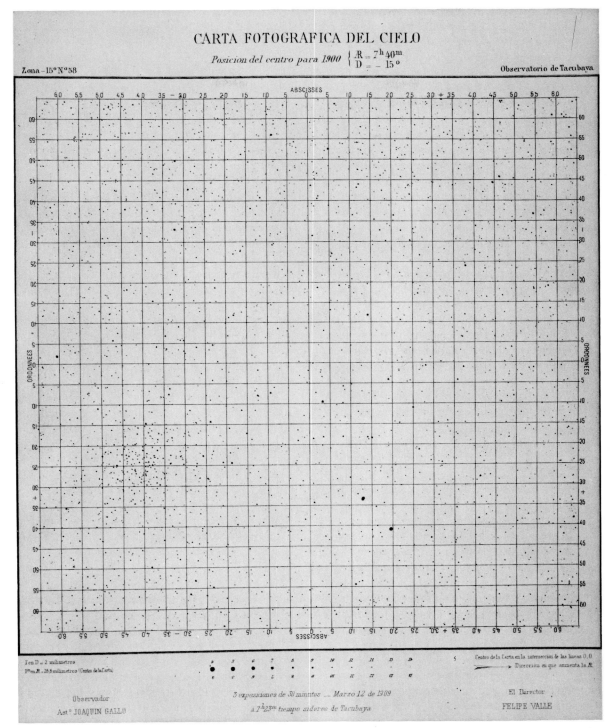

2.7. A *Carte du Ciel* plate made at the observatory of Tacubaya in Mexico in 1909. The field is two degrees square. South is at the top. Note the open cluster, M 46, at the lower left.

reasons. As Holden, the first director of the Lick Observatory, wrote, "We should hardly be willing to bind ourselves to a programme which exacted so much routine work for so long." As an observatory director, Holden wanted more flexibility than participation in the *Carte du Ciel* would allow. Other directors like Pickering at Harvard were moving into astrophysics. They needed funds to support research in solar physics, photometry and spectroscopy. Clearly, involvement in the *Carte* would work against these goals, for participation would entail the resources of even the wealthiest American observatories for the foreseeable future. By a curious turn of events, then, the growth of astrophysics in the United States may have been stimulated as a consequence of non-participation in the *Carte du Ciel* while, at least to a degree, astrophysical research in Europe may have been retarded because the *Carte* absorbed funds and engrossed staff time that otherwise might have been allocated to astrophysics.

Lack of American participation notwithstanding, the Astrographic Congress and the *Carte du Ciel* ushered in a new epoch in the history of astronomical photography. The Permanent International Committee educated astronomers to the value of photography, and resistance to the new research technology declined.

Acceptance of photographic reflectors by professional astronomers

Photographic research with reflecting telescopes remained the domain of amateurs until the closing years of the nineteenth century. So widespread was professional distrust of reflectors that even after silver-on-glass mirrors replaced those made of speculum metal few professionals experimented with them. During the discussions by national commissions preparing for the 1874 transit of Venus, for example, German, French and British professionals all rejected the reflector because they considered it virtually impossible to keep in optical adjustment and extremely difficult to manipulate. P.J.C. Janssen, founder of the Meudon Astrophysical Observatory, was something of a pioneer when he equipped the new institution with a 1-m, f/3 Newtonian reflector designed for photographic spectroscopy of nebulae and comets.

J.E. Keeler (1857–1900), second director of the Lick Observatory, and G.W. Ritchey (1864–1945),

a gifted optical engineer associated with the Yerkes and later Mt Wilson observatories, finally convinced sceptical professionals of the reflector's capabilities. After Keeler assumed the directorship of Lick in 1898, he devoted himself to modernizing the Crossley 36-inch (91-cm) reflector (f/5.8). This instrument, a gift to the observatory in 1895, had resisted all previous attempts by Lick astronomers to bring it into service. After months of painstaking labour, Keeler began a programme of nebular photography, and in the two years remaining to him he published more than twenty papers reporting his photographic studies of these objects. On the basis of photographic evidence, Keeler reached three important conclusions. "Many thousands of unrecorded nebulae exist", he wrote in *The Astrophysical Journal* in 1900, and "a conservative estimate places the number within reach of the Crossley reflector at about 120000." Nebulae "exhibit all gradations of apparent size, from the great nebula in *Andromeda* down to an object which is hardly distinguishable from a faint star disk". Further, "most of these nebulae have a spiral structure". Keeler's photographs helped open a new field of astrophysical research.

In 1901 Ritchey described a 24-inch (61-cm) reflector (f/4 in the Newtonian configuration and f/19 when used as a Cassegrain) constructed in the Yerkes shops. It was designed specifically for photography. Ritchey and his team gave special attention to perfecting the optical components, the stability of the mirror supports and the rigidity of the skeleton tube. The goal was an optical train that would remain in perfect adjustment during exposures of five or six hours. With justifiable pride the Yerkes astronomer concluded in 1904, "In nearly all respects the same degree of care and refinement was used in the making of this instrument as is given in the case of the best modern refractors." Ritchey went on to compare reflectors and refractors as photographic instruments. He stressed the advantages that perfect achromatism gave the reflector and pointed out that while lenses do not transmit all the light they receive, a carefully silvered mirror wastes very little.

The Yerkes reflector produced extraordinary photographs that far surpassed those Keeler made with the Crossley. The 24-inch covered a larger field and its optical perfection and mechanical stability resulted in brighter and sharper images.

Soon, the Crossley was remounted in a form that permitted Lick astronomers to realize its full potential.

Max Wolf (1863–1932), director of the Königstuhl Observatory at Heidelberg and a leader in the application of photography to research on the Milky Way and extragalactic objects, must have agreed with Ritchey's conclusion that photographs taken with a reflector "afford striking illustrations of the wonderful richness and complexity of structure of the nebulae, and of the importance of photography in the study of these very faint objects". By 1906 Wolf had installed the Waltz 71-cm reflector and was using it for photographic studies of nebulae.

By 1908, when Ritchey completed the 60-inch (152-cm) reflector at Mt Wilson, improvements in the engineering and optical design of reflecting telescopes had caused professional astronomers to revise their earlier opinions. Large, carefully constructed photographic reflectors became indispensable tools for astrophysical research. As the nineteenth century came to an end, photography found numerous applications in such research fields as solar physics, spectroscopy, solar system investigations and the study of the Milky Way and extragalactic objects, all of which are dealt with elsewhere in this volume. The concluding portion of this chapter examines the role of photography in the study of astrometry and stellar photometry.

Photographic astrometry

Astrometry includes the study of the places of the stars based on a uniform system of coordinates and data on their proper motion, and, where it can be determined, stellar parallax. In astrometric research, standards of precision are extremely high and accuracy of one micron (0.001 mm) or better is the rule. Work at this level of precision involves a continual search for errors inherent in instruments as well as those introduced by observers. Detection of such minute errors demands that workers in astrometry possess a special mix of talents, including practical engineering skills and a knowledge of applied mathematics and statistics. Furthermore, a successful investigator must be able to organize a research programme that will produce significant quantities of data meeting basic standards of precision in an efficient and economical manner. Here astronomers share the

goals of industrial capitalists in seeking to introduce economies of scale and employ methods that will make the enterprise cost effective. Photography proved especially useful for the mass production of precise data.

The first convincing determination of stellar parallax, that of the double star 61 Cygni, occurred in 1838. Until the 1880s the work of measuring stellar parallax was generally carried out with the heliometer, a split objective refractor developed by Fraunhofer and capable of measuring very small angles precisely. When the Oxford astronomer C. Pritchard (1808–93) started experimenting with photography for parallax measurement, the distance of approximately ninety stars had been determined.

Pritchard was one of the most remarkable figures in Victorian astronomy. Elected Savilian Professor at Oxford in his sixty-second year, the former schoolmaster had served as president of the Royal Astronomical Society but had no significant astronomical investigations to his credit. In spite of this inauspicious background, Pritchard embarked on an ambitious programme of innovative research. A recent biographer has remarked that "the most important aspect of Pritchard's work at Oxford was his role in convincing the astronomical community that accurate measurements of precision could be obtained from photographic plates". Using the 13-inch (33-cm) reflector built by De la Rue and later presented to Oxford, 330 plates of 61 Cygni were exposed on eighty-nine different nights during a period of thirteen months beginning in May 1886. These plates, 5 cm square, were subjected to over 30000 measurements. In his annual report to the Board of Visitors in 1887 Pritchard announced, "The somewhat hazardous enterprise of attempting, for the first time in the history of astronomy, to obtain the distance of the fixed stars from our Earth by the aid of photography has been attended with success." His enthusiastic review concluded: "Astronomical photography is hereby placed on a secure basis as an efficient and exact exponent of the highest form of astronomical science [astrometry]."

During the 1890s the collection of astrometric data given by L. M. Rutherfurd to the Columbia University Observatory was being utilized by H. Jacoby and J. K. Reese and their graduate students. Indeed, it was the Columbia-trained astronomer F.

Schlesinger (1871–1943) who defined the state of the art for photographic measurement of stellar parallax. As a staff member at Yerkes Observatory, Schlesinger developed the methods and techniques that became standard for the photographic determination of stellar parallax using long-focus refractors. His methods and preliminary findings were presented in 1910 and 1911 in a series of papers in *The Astrophysical Journal*. After becoming director of the Allegheny Observatory in 1905, Schlesinger designed the first long-focus photographic refractor (objective 30 inches (76 cm), focal length 46 ft (14 m)) specifically for the study of stellar distances.

The photographic methods pioneered by Schlesinger at Yerkes included the following basic operations and techniques. He decided after extensive experiments that for exposures no longer than five minutes using isochromatic plates whose maximum sensitivity lay between 5200 and 5700 Å, good focus could be achieved using a visual refractor without filters. (Schlesinger feared distortion introduced by a filter.) At Yerkes he employed 20×25-cm plates that covered a field of 35×45 arc minutes. A major source of error in photographic astrometry stems from differences of more than one magnitude between the star whose parallax is sought and comparison stars. Schlesinger conducted numerous trials seeking to reduce the light of the parallax star. Finally he tried using a rotating sector driven by a small electric motor. Depending on the size of the adjustable aperture, the light of the parallax star could be reduced up to seven magnitudes. In order to control errors introduced by differential flexure of the tube and objective, all plates were exposed with the telescope on the west side of the pier. Schlesinger also concluded that it was desirable to maintain the same hour angle for parallax plates, and that parallax photography could most fruitfully be accomplished by dividing the night into two shifts: from the end of astronomical twilight until about 10 p.m. local time and then again from about 2 a.m. until the beginning of astronomical twilight forced the observer to close the dome.

Once successful methods for the photographic determination of stellar parallax had been developed, Schlesinger turned his attention to the use of photography for the precise measurement of star positions. By 1916 he was able to report on the construction and use of an f/21 doublet (effective aperture 3 inches (8 cm)). Using special flat plates 20×25 cm and 3 mm thick, the instrument covered a field of 25 square degrees. The great size of the field permitted inclusion of a large number of standard reference stars whose positions had been accurately determined by visual observers using transit circles. This led to a significant gain in accuracy in contrast to the *Carte du Ciel* plates, which covered only 4 square degrees and for which it was often difficult to find suitable reference stars. Satisfied that he had located and corrected systematic errors in the instrument, Schlesinger suggested that the reobservation of the Astronomische Gesellschaft (AG) zones (the original project, carried out visually, involved sixteen observatories and spanned half a century; see Volume 3) could now be accomplished by a single observatory using a photographic doublet. He argued that in contrast to a probable error of well over $0''.50$ in both right ascension and declination in the original *AG* observations, the photographic doublet would reduce the probable error for any one observation to $0''.18$. Later, German astronomers would use photography in preparing the *AGK2* and *AGK3*, that is, the second and third Astronomische Gesellschaft catalogues, which set new standards for precise star positions.

Photographic photometry

Astronomers seeking to apply photography to the determination of stellar magnitudes faced problems analogous to those that confronted workers in photographic astrometry. Various systematic errors had to be discovered and taken into account, and procedures standardized and carefully followed. Several critical issues required international agreement if a homogeneous scale embracing both bright and faint stars was to be developed. Only following a lengthy period of experimentation, discussion and negotiation did the international astronomical community reach consensus on these matters.

Until about 1900 research in photographic photometry went forward at an uneven pace. On both sides of the Atlantic observatories were at work on photometric problems, but astronomers not associated with the *Carte du Ciel* had little opportunity to communicate with each other. Lack of agreement on the physics and chemistry of the photographic process proved a handicap. Many astronomers continued to assume that the action of light on the photographic plate was independent of its wavelength, while the applicability of the Bunsen–Roscoe law of reciprocity to astronomical photography was still being asserted as late as 1912. (The Bunsen–Roscoe law of photographic photometry states that the darkening of an image on a photographic plate depends simply on the total amount of light that strikes the grains of silver halide; thus, if the intensity of the light is cut in half the exposure time must be doubled.)

Differences of approach between workers in the field also caused difficulties. Astronomers associated with the *Carte du Ciel* approached the photographic determination of stellar magnitude from a statistical point of view. Younger astrophysicists like K. Schwarzschild in Europe and E. C. Pickering in the United States, who were not committed to the *Carte du Ciel* project, found it more fruitful to approach the photographic determination of magnitude as a physical problem.

The 1887 Astrographic Congress was forced to postpone a decision on standards and methods for the determination of photographic magnitudes. There simply was not enough information at hand. In 1889 the Permanent International Committee took up these questions. H. G. v. d. S. Bakhuyzen, director of the Leyden Observatory, proposed a numerical approach. B. A. Gould, first director of the Cordoba Observatory, developed a formula for estimating the number of stars down to a specified magnitude that should be found in a given area of the sky. Bakhuyzen suggested the formula be used to predict the number of stars down to the eleventh magnitude that ought to appear on a plate covering four square degrees; thus observers had only to count the stars registered using various exposures until they reached the predicted value. Prosper Henry countered the Bakhuyzen proposal: assuming the accuracy of the Bunsen–Roscoe law, the Paris astronomer suggested that the Committee ought to think in terms of Pogson's ratio (two stars that differ by one magnitude have a brightness ratio of $^5\sqrt{100}:1$ or 2.512) and so increase the exposure time needed to reach stars known to be of the ninth visual magnitude by a factor of 2.5^2 (or 6.25) in order to record stars of the eleventh. Of course, Henry and most other delegates were assuming a

virtual identity between visual and photographic scales.

The 1889 meeting adopted Henry's proposal, but in 1891 the matter was reopened and the Permanent International Committee continued its discussions. Doubts were raised concerning the validity of the Bunsen–Roscoe law for astronomical photography: perhaps the blackening of the stellar image was not directly proportional to the length of exposure. It was believed, however, that the diameter of a star image varied proportionally and that determination of the size of star images would be a more satisfactory gauge of magnitude. It would be easy enough to measure the diameter of stars whose magnitudes were given by Argelander or other authorities; astronomers could then develop equations expressing the relationship between exposure and diameter of stars for known magnitudes and extrapolate to fainter objects.

Kapteyn urged the Committee to approve an alternative plan involving the use of wire mesh screens to reduce the light of stars by a known factor (in this case, two magnitudes), so that the second order images of a ninth-magnitude star could be taken as representing magnitude eleven. But while none of the delegates quarrelled with the physical theory behind the use of wire screen diffraction gratings (reduction of brightness is a function of the thickness of the wires and the space between them), they were unenthusiastic.

Consensus on these issues proved so elusive that in 1896 the Permanent International Committee shifted the burden of defining photographic magnitudes to the several observatories taking part in the *Carte du Ciel* project. Resolution 6 of the 1896 meeting granted participating institutions the right to determine photographic magnitudes by measurement or estimate, but the Committee declined to specify guidelines or procedures to be followed. Instead, it requested that each observatory employ an approach that could be explained in precise detail so that the various magnitude scales could be reduced to a common system.

Virtually all observatories participating in the *Carte du Ciel* employed the size of the star image as the basis for determining photographic magnitudes. Astronomers developed empirical formulae expressing the relationship between image size D and magnitude m. These calculations were based on established visual magnitudes and involved using the statistical procedure known as least squares to determine the constants a and b for each plate. By the end of the century a variety of formulae appeared in the literature. The following examples illustrate the range:

Christie (Greenwich) $m = a + 2.5 \log(t - bD)$, where $t =$ length of exposure

Kapteyn (*Cape Photographic Durchmusterung*) $m = a/(b + D)$

Scheiner (Potsdam) $m = a + bD$

Turner (Oxford) $m = a - bD$

The differences between these equations can be explained in part by the optical characteristics of the various astrographic refractors and different photographic practices employed. The sensitivity of the plates and local atmospheric conditions also played a role. However, as Abney remarked in 1891, "Anybody who has attempted to measure the disk of a star will know that it is rather a shaky thing to do. It is very difficult to say where the disk extends to and where it does not."

A fresh approach to photographic photometry emerged when K. Schwarzschild (1873–1916), an astronomer at K. v. Kuffner's private observatory in suburban Vienna, suggested that diffraction seriously affected the size of star images at varying distances from the optical axis, and for this reason the measurement of image diameter could not provide an accurate gauge of magnitude. Instead, he proposed to measure the density of extrafocal images. While others had speculated concerning this technique, apparently Schwarzschild was the first to employ it. Furthermore, Schwarzschild undertook laboratory studies of the Bunsen–Roscoe law and was able to quantify the failure of reciprocity that occurs at low levels of illumination. In 1900 Schwarzschild reported in *The Astrophysical Journal*: "The following rule . . . should replace the [Bunsen–Roscoe] law of reciprocity: Sources of light of different intensity I cause the same degree of blackening under different exposures t if the products $I \times t^{0.86}$ are equal."

Moving to Göttingen in 1901, Schwarzschild continued research in photographic photometry. Equipped with a Zeiss camera (45-mm aperture and 460-mm focal length) and using 16×21-cm plates, he was able to photograph a field of twenty degrees declination and one full hour in right ascension at the Equator. Schwarzschild's plan involved determining the photographic magnitudes of approximately 3500 BD stars down to about visual magnitude 7.5 in a zone from $0°$ to $+20°$.

Schwarzschild equipped the Göttingen instrument with a mechanism called a Schraffierkassette, which moved the plateholder in a zigzag fashion in order to build up square star images (slightly out of focus) 0.25 mm on a side. The time needed to complete a full run of the Schraffierkassette was 3 minutes and 45 seconds. Image intensity was then measured with a newly-devised Hartmann microphotometer.

Three exposures were made on each plate in the ratios $1:3:9$, resulting in a difference of approximately one magnitude between consecutive images of a given star. Because the plates overlapped, Schwarzschild and his assistants were able to work from one plate to the next, carrying the scale of magnitude around the zone in both directions until the two series met. This permitted an accurate determination of systematic errors. Schwarzschild adopted for the zero point (where visual and photographic magnitudes are equal) type A stars of magnitude 6.3 based on the visual photometry of Müller and Kempf. He also developed the concept of colour indices (photographic *minus* visual magnitudes) as part of this project. Schwarzschild's work was widely discussed and the two volumes of the *Göttingen Aktinometrie* did much to convince astronomers of the value of a physical rather than statistical approach to photographic stellar photometry.

Schwarzschild's concept of colour index (a negative value for hot stars of the O and B spectral types and positive for cool red stars from A1 to M) became increasingly significant for twentieth-century astrophysics. Colour indices are closely related to stellar temperature and luminosity, and when found to be in excess of predicted values provide information on the interstellar medium. Because they are relatively easy to measure, colour indices are frequently used in graphic presentations such as the Hertzsprung–Russell diagram in which absolute magnitude is plotted against colour index.

E. C. Pickering, who served as professor of physics at the Massachusetts Institute of Technology (MIT) before assuming the directorship of the Harvard College Observatory, reported his earliest experiments with photographic photometry at the June 1883 meeting of the Royal

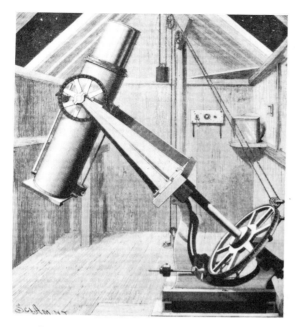

2.8. The Bache 8-inch photographic refractor at Harvard College Observatory.

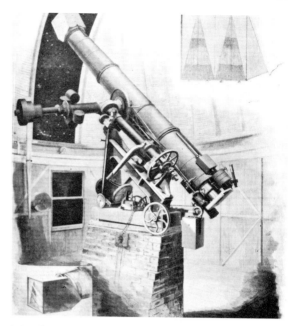

2.9. The Draper 11-inch photographic refractor with objective prism at Harvard College Observatory.

Astronomical Society. Only at the end of the 1880s, however, did members of the Harvard staff begin sustained research in photographic photometry. The director's brother, W.H. Pickering (1858–1938), who had previously taught courses in photography at MIT, was placed in charge of photographic research. He was soon joined by E.S. King (1861–1931). Henrietta S. Leavitt (1868–1921) was in charge of measuring photometric plates. Over the years Miss Leavitt assumed more and more responsibility for discussing the data as well.

Miss Leavitt's career exemplifies one of the unanticipated consequences that followed the introduction of photography into astronomical research. Collections of photographic plates grew rapidly and observatories had to add new staff members in order to keep up with the influx of data. As early as the 1860s Rutherfurd hired women to measure plates. By the 1880s Harvard was employing women in that capacity. They measured, frequently reduced and sometimes discussed photographic data.

Photographic photometry at Harvard can be divided into several phases. Harvard astronomers early on rejected the measurement of stellar diameter as a means of establishing magnitudes. As W.H. Pickering reported in 1895 in *Annals of*

Harvard College Observatory, "We measure magnitudes by comparison with a photographic scale . . . and we adopt the visual magnitudes of average first type stars [Secchi's white or blue stars] as standards, making the photographic magnitudes conform to them, instead of taking a certain star as a standard and assuming that a star whose image is equally distinct with two and a half times the exposure is one magnitude fainter." The first experiments with photographic scales were carried out in 1888 when the Bache telescope was used to photograph the Hyades with exposures of 729, 243, 81, 27, 9 and 3 seconds. Portions of these plates were cut out, mounted and used for direct comparison with stars of unknown brightness. Catalogues were soon published listing photographic magnitudes of 1009 stars within one degree of the North Pole, 420 stars in the Pleiades and 1131 stars near the Equator.

By the late 1890s King was deeply involved in a variety of experiments dealing with the properties of the photographic plate. Starting in 1900 he sought to determine the photographic magnitudes of bright stars using extra-focal images. Like Schwarzschild, King wanted to develop a scale of absolute magnitudes that he defined in 1912 in the Harvard *Annals* as *independent of any other series or system of magnitudes, visual or photographic. All the data for determining these relations have been derived from the plates used in this investigation. The only point of contact with the photometric magnitudes [visual magnitudes published in volume 50 of the* Annals] *has been the condition, that the mean of the photographic magnitudes for stars of Class A should agree with the photometric values.*

In 1907 E. C. Pickering announced plans for determining the photographic magnitudes of a sequence of stars in the immediate vicinity of the Pole. When photographed on the same plate as a field containing stars of unknown brightness, the North Pole Sequence would provide a standard. Miss Leavitt was entrusted with the investigation. The North Polar Sequence included ninety-six stars ranging from the fourth to twenty-first magnitudes. Published in 1917 in volume 71 of the Harvard *Annals*, the sequence rested on 299 plates taken with thirteen telescopes whose apertures ranged from half an inch (1 cm) to the 60-inch (152-cm) Mt Wilson reflector. Numerous techniques were used to vary the light of the stars, including plates of Iceland spar, wire screens and diaphragms, as well as use of extra-focal images. The range of instruments and variety of techniques were intended to eliminate systematic errors. The final results were based on a statistical analysis of the data reduced to a standard zero point; however, determination of the colours of faint stars introduced uncertainties.

For many years Pickering served as a member of the commission on photographic magnitudes appointed by the Permanent International Committee of the *Carte du Ciel*. He was often frustrated by the difficulties encountered in trying to reach agreement on key issues. Finally, in December 1910, Pickering was able to write to the director of the Paris Observatory that the commission, whose members had consulted frequently during the 1910 meeting of the International Solar Union held in California, had agreed on a definition of the zero point put forward by Schwarzschild. The international standard equated visual and photographic magnitudes of Class A stars between magnitudes 5.5 and 6.5 in the Harvard visual photometry.

Early in the new century a Yerkes Observatory team led by J. A. Parkhurst (1861–1925) joined the ranks of astronomers employing a physical approach to photographic photometry. Parkhurst, whose interest in stellar photometry went back to his days as an amateur variable-star observer and volunteer at the newly-organized Yerkes Observatory in the 1890s, began his research with extensive laboratory investigations. On the basis of this work a set of carefully calibrated sensitometer images was added to each plate after exposure to the sky. These images formed the basis for a scale of 'absolute' magnitudes (that is, magnitudes independent of other information or assumptions).

Unlike previous investigators, Parkhurst published extensively while his research was in progress. In a series of papers that appeared in *The Astrophysical Journal* he meticulously analysed sources of error and elaborated on precautions necessary for successful work in photographic photometry. As each plate was measured, corrections were applied for the distance of the image from the optical centre, atmospheric absorption, and colour-error of the objective. The "Yerkes Actinometry" was published in 1912. It contained photographic and photovisual magnitudes, colour indices and spectral types for the stars in Müller and Kempf's visual photometric Durchmusterung from $+73°$ to the Pole.

At Greenwich, new photometric methods were introduced by S. Chapman (1888–1970) and P. J. Melotte (1880–1961) in order to standardize photographic magnitudes in preparation for their monograph, "The numbers of stars of each photographic magnitude down to $17^m.0$ in different galactic latitudes", published in *Memoirs of the Royal Astronomical Society* in 1915. These young scientists redetermined the photographic magnitude of 262 stars within 25 arc minutes of the Pole using wire diffraction gratings mounted in front of the objective of the 26-inch (66-cm) refractor. They had found diffraction gratings especially valuable for work with faint stars, and their discussion of the construction and theory of wire diffraction gratings in 1913 was one of the first to be published.

F. H. Seares (1873–1964), using the 60-inch (152-cm) reflector at Mt Wilson, perfected photometric techniques for large reflecting telescopes. Seares, whose interest in photometry had developed while director of the Laws Observatory at the University of Missouri, joined the Mt Wilson staff in 1909. He brought to photometric research a deep concern for precision and scrupulous attention to the details of observing acquired as a student of astrometry and celestial mechanics at the University of California at Berkeley. Seares attempted to create a uniform scale of magnitude encompassing bright (to tenth magnitude), intermediate and faint stars (below magnitude 18). To achieve this goal he had to reduce the aperture of the reflector for bright objects, and to wait for exceptional nights in order to photograph faint stars. Circular diaphragms as well as single and double thickness wire mesh screen were employed. On the basis of work in the laboratory and at the telescope, Seares demonstrated the reliability of diffraction screens. His findings indicated errors no larger than two- to three-hundredths of a magnitude over a five-magnitude range. In 1922 Seares became the first chairman of the Commission on Stellar Photometry of the International Astronomical Union. The IAU adopted his values for the North Polar Sequence as international standards.

By the 1922 meeting of the IAU Commission 23, astronomers could look back on a quarter-century of rapid development during which photographic photometry solved a number of problems relating to theory, instrumentation, methods and techniques. The international astronomical community were agreed upon fundamental standards, including a zero point and provisional values for the North Polar Sequence. Consensus had been reached by treating the photographic determination of magnitude as a physical problem.

The diffusion of photographic technology through the international astronomical community was accomplished in approximately three generations. The process was closely linked to the rise of astrophysics and involved the activities of both amateur and professional astronomers. The *Carte du Ciel* and the Permanent International Committee established to supervise the project played a key role in the diffusion process. By 1920 photography had become a major research tool in most areas of astronomical investigation. Even at that date, however, photography proved of little value in transit circle work, for the study of double stars and in mapping lunar and planetary surfaces. Nor had astronomers yet succeeded in photographing the corona out of eclipse. But in time even these fields would make use of the photographic plate.

Further reading

D. Hoffleit, *Some Firsts in Astronomical Photography* (Cambridge, Mass., 1950)

John Lankford, Amateurs and astrophysics: a neglected aspect in the development of a scientific specialty, *Social Studies of Science*, vol. 11 (1981), 275–303

John Lankford, Photography and the long-focus visual refractor: three American case-studies, 1885–1914, *Journal for the History of Astronomy*, vol. 14 (1983), 77–91

Daniel Norman, The development of astronomical photography, *Osiris*, vol. 5 (1938), 560–94

G. Rayet, Notes sur l'histoire de la photographie astronomique, *Bulletin astronomique*, vol. 4 (1887), 165–76, 262–72, 307–20, 344–60, 449–56

J. Scheiner, *Die Photographie der Gestirne* (Leipzig, 1897)

H. H. Turner, *The Great Star Map: Being a Brief General Account of the International Project Known as the Astrographic Chart* (London, 1912)

Gérard de Vaucouleurs, *Astronomical Photography: From the Daguerreotype to the Electron Camera* (London, 1961)

Harold F. Weaver, The development of astronomical photometry, *Popular Astronomy*, vol. 54 (1946), 211–30, 287–99, 339–51, 389–404, 451–64, 504–26

3

Telescope building, 1850–1900

ALBERT VAN HELDEN

Introduction

The period 1850–1900 saw the growth of the refracting telescope to the greatest size it has attained to date, and also the introduction of the modern reflecting telescope with its metal-coated glass mirror. In the drive for ever greater light-gathering power, astronomers were faced with the choice between refractors and reflectors. The advocates of the two types of instrument carried on a lively debate throughout this period against a background of changing astronomical practice. The new research tools, spectroscopy and photography, and the increasing emphasis on stars and nebulae favoured the reflector in the long run. But during this period the refractor remained the preferred instrument among professional astronomers, while the cause of the reflector was furthered by wealthy amateurs.

In 1850 the largest refractors were the 15-inch (38-cm) instruments at Pulkovo and Harvard, both made by the firm of Merz & Mahler in Munich. The largest reflector was the 72-inch (182-cm) 'Leviathan' erected in 1845 by William Parsons, third Earl of Rosse, at Birr Castle in Ireland. A comparison of these instruments shows the former to be the epitome of instrumental refinement, true precision instruments, while the latter was a crude exercise in raw power. It is not difficult to see why refractors were preferred for the precision work of observatories.

The refractor was a more sturdy and precise instrument, eminently suited for traditional position measurements. Beyond periodic cleaning, the optical system needed virtually no attention. Its greatest drawbacks were residual chromatic aberration, the limitation on aperture size caused by the difficulty of obtaining large glass blanks for the objectives, and high cost. The reflector had the advantages of the lack of chromatic aberration, a much greater possible aperture and a lower cost because only one primary optical surface had to be figured. Part of this advantage was lost however by the rapid tarnishing of the copper–tin alloy known as 'speculum metal'. The mirror had, therefore, to be repolished frequently, and this was always a major undertaking without guaranteed success. Only after the introduction of silvered-glass mirrors could the reflector seriously compete with the refractor. The reflector was difficult to manoeuvre and did not lend itself easily to measurements. Because of its superior light-gathering power but lack of precision it served best in the area of discovery, especially in the realm of the nebulae.

Refractors

Glass-making

The limiting factor in the quest for greater light-gathering power with the refractor was, at any point in time, the difficulty of producing glass blanks of the requisite size, free of impurities and sufficiently homogeneous for optical purposes. Flint glass (made from sand, potash and lead oxide) was more troublesome in this respect than the more ordinary crown glass (made from sand, potash and lime), because the lead oxide is much heavier than the other ingredients and therefore tends to segregate towards the bottom of the crucible. In the first decade of the century P.L. Guinand (1748–1824) had solved this problem. Continuous stirring during the cooling of the mixture made the glass homogeneous and eliminated bubbles, and the use of burned clay for the crucible and the stirrer (in the form of a hollow cylinder) eliminated contamination.

At the Munich Optical Institute, Guinand had taught his method to J. Fraunhofer, and before his untimely death in 1826 Fraunhofer had passed the secret on to G. Merz and F.J. Mahler. Merz and Mahler bought the Institute in 1839. Even after their deaths the firm of Merz & Mahler remained in the forefront of glass-making and lens grinding for many years.

In France, Guinand's son Henri sold his father's secret to G. Bontemps who, starting in 1828, produced good optical glass. In 1837 Bontemps sold the secret to the Chance Brothers firm in Birmingham, and L. Chance took out a patent on the process the following year; when this patent was printed in 1857, the process became generally known. In the meantime, Bontemps himself moved to England after the 1848 revolution, and following his affiliation with Chance Brothers this firm became one of the few suppliers of large blanks for objectives. Upon his return to France, Bontemps published Guinand's method of producing optical glass (*Guide du Verrier*, 1868).

Henri Guinand's grandson, C. Feil, carried on the glass-making tradition of his family in France. Before Feil's death in 1887 the firm Feil-Mantois made glass blanks for the objectives of a number of large telescopes. Chance Brothers and Feil-Mantois produced virtually all the blanks for the very large refractors of the period 1850–1900.

Guinand's process for both crown and flint glass was as follows. A burned-clay crucible, about a metre high, cylindrical in form and tapered towards the top, was heated in a kiln to white heat. The materials were then added slowly to prevent boiling, after which the kiln was completely blocked. The temperature was now raised to 1600–1800 °C, and maintained for twenty to thirty hours. Small samples were then taken, cooled in round moulds, and examined under a magnifying glass using light of various colours. If this examination revealed no bubbles, the fire was gradually lowered and the mixture skimmed to remove impurities.

The molten glass was now ready for mixing. The clay mixer was attached to an iron bar with a hook, by means of which it was manipulated. The stirrer was brought to white heat before it was introduced into the melt. The stirring was usually done in five-minute relays by a team of men. If bubbles appeared during the first hour of stirring, the kiln was reblocked and the temperature kept high until no further bubbles appeared. In the absence of bubbles, stirring was continued for about fifteen hours while the temperature was lowered very gradually. After this time the glass had become too viscous to allow further agitation, and the stirrer was removed and the fire completely extinguished.

The 600 to 1000-kg melt was now allowed to cool more rapidly for about six hours. When the glass reached the temperature at which it gave a sharp report when tapped with a metal implement, the kiln was completely reblocked to protect the glass from air currents. The glass was now cooled very slowly, the process taking from several days to several months, depending on the type of glass and the size of the blanks desired. The crucible was now broken and the glass (in one or more lumps) was examined for impurities. These impurities were hammered out, and the piece or pieces were placed in appropriately shaped moulds, placed in a furnace, and heated to 800–900 °C. The glass now flowed into the shape of the mould. After moulding the glass was reheated and cooled slowly to relieve stresses. All being well, the result was a large glass blank of high optical quality. Although the manufacture of optical glass was a profitable business for the few firms engaged in it, the risks were great when it came to producing very large blanks for refractors.

If the Guinand process, which had become generally known by the 1860s, solved the problem of making large batches of homogeneous glass, free of impurities, the range of glasses available, and control over their optical properties, were still very limited. For making objectives, opticians had to be satisfied with the few grades of crown and flint glass which the glass manufacturers could supply. But crown and flint glass have different partial dispersions: that is, they create spectra which are internally differently proportioned. Chromatic aberration, therefore, could be corrected only over a certain part of the spectrum by the use of a combination of crown and flint. At the extremes of the spectrum objectives were over- and under-corrected. This condition, the so-called 'secondary spectrum', was a matter of continuing concern to astronomers, and its importance increased with the introduction of the new tools of photography and spectroscopy. Given the fact that chemical composition affects the optical properties of a glass, was it possible to create new glasses with more desirable optical properties?

As early as 1824, upon the request of the Royal Society, Michael Faraday, John Herschel and George Dollond combined in their efforts to improve optical glass. But the experimental melts made by Faraday were not successful and the effort was abandoned in 1830. A few years later, W. V. Harcourt began a long series of experiments, which were restricted mostly to phosphate glasses combined with fluorides and sometimes other compounds. The results of the 166 melts were tested by G. Stokes, but Stokes had great difficulties with these tests, and the results he announced in the 1870s were disappointing. They did, however, hold out enough hope for J. Hopkinson, head of the optical department of Chance Brothers, to pursue a course of experiments extended over a year. But Hopkinson's results, announced in 1875, were likewise inconclusive. By that year, however, the leadership in these experiments had already passed to Germany.

Carl Zeiss (1816–88), who had been trained as an instrument maker, opened a shop in the German town of Jena in 1846, and was appointed instrument maker for the local university in 1860. Zeiss specialized in microscopes and wished to base the design of these instruments on scientific principles rather than trial and error. In this effort he was successful after 1866, when he interested E. Abbe (1840–1905), a lecturer in physics at the university, in this problem. Over the next decade Abbe revolutionized microscope design and the Zeiss firm (in which Abbe became a partner in 1876) grew enormously.

But Abbe was limited in his designs by the optical properties of available glasses. He was entirely confident that glasses with a much wider range of optical properties could be made, but glass makers, including Feil in Paris, could not satisfy his requests. In the late 1870s Abbe came into contact with F. O. Schott (1851–1935), a young glass chemist who was having some success in this area, and in 1884 Zeiss, Abbe and Schott founded Schott & Associates in Jena, a firm which was to provide the Zeiss Company with optical glass. These two companies working in close cooperation were the model for the modern scientific glass industry. The results came quickly, and by 1886 Schott was offering for sale no fewer than forty-four types of optical glass with a wide range of properties. Optical design was now freed from the limitations

imposed hitherto by the narrow range of the properties of glass. By the turn of the twentieth century the new types of glass had become established in microscopes, and telescopes of modest apertures.

Lens-shapes
Thanks to the work of Fraunhofer on spectral lines, refractive indices and partial dispersions of glasses could be measured very accurately. Once these properties were known for the particular crown- and flint-glass discs, the curvature of the components that were to make up the objective could be calculated. Ideally, the combination should eliminate chromatic aberration and spherical aberration over the entire spectrum, while also correcting the extra-axial defects of coma and oblique astigmatism. In practice, this was impossible: defects could be minimized only over a specified range of the spectrum and a given extent of the field of view.

Chromatic aberration was minimized between the C and F lines of hydrogen, the region of the spectrum to which the human eye is most sensitive. If, however, the objective was to be used for photographic purposes, correction in the violet range became crucial, for the greatest sensitivity of early photographic plates lay in this region. If the telescope was to be used for both visual and photographic purposes its objective had to be a specially designed triplet or the instrument had to be equipped with a removable photographic corrector. The refractors made for the cooperative *Carte du Ciel* effort had entirely separate visual and photographic optical systems.

Spherical aberration was not corrected by making the curvature of the objective non-spherical (in the theoretical manner already outlined by Descartes). Spherical curvatures were used throughout the nineteenth century, and the curvature of each component was chosen so as to correct spherical aberration at several wavelengths. Whereas Clairaut had been able to correct it only at one wavelength, Gauss showed how to correct it at two wavelengths in a doublet.

Coma, the defect that makes images away from the optical axis comet-shaped, was corrected first by Fraunhofer, but his method was not made public. Abbe formalized the condition for the suppression of this defect: if through the doublet the incident rays have a net refraction in proportion to their distance from the optical axis (the so-called sine-condition) the system will have the same focal length for any portion of the aperture and coma will be absent.

The simultaneous minimization of the various optical defects was a complex problem. The mathematician was confronted with the optical characteristics of the crown and flint blanks available, and the desired aperture and focal length of the objective; to these were added a number of other conditions that determined the particular solution arrived at. The purpose of the telescope was crucial. For qualitative visual purposes the optical defects had to be minimized over a very small range around the optical axis only, so that extra-axial defects were not a great drawback. But if angular separations of stars were to be measured with the instruments, coma had to be minimized. For photographic purposes defects further removed from the optical axis also had to be considered. The range in which chromatic aberration had to be suppressed depended on whether the instrument was to be used for visual or photographic purposes or both. But there were also constructional considerations. Although the number of defects that could be corrected depended on the number of components of the objective, triplets were used infrequently and doublets were usually preferred. The crown component was almost always put in front because of this glass's superior hardness and resistance to chemical action. Curvature was kept at a minimum for several reasons, not the least of which was the difficulty of centring two highly curved components. Often rather arbitrary conditions were added, as for example, of making the curvatures of the two faces of the crown lens equal. In small objectives the curvature of the rear face of the crown was often made equal to that of the front face of the flint so that the components could be joined together by means of Canada balsam.

The complicated computations taxed the skills of the finest mathematicians. Gauss based his solutions on the first term of the MacLaurin expansion $\sin \theta = \theta - \dfrac{\theta^3}{3!} + \dfrac{\theta^5}{5!} - \dfrac{\theta^7}{7!} \ldots$, which sufficed for very small angles of incidence. Over a larger field of view, however, the third-, fifth- and even seventh-order terms could not be neglected. In 1855 P. L. v. Seidel extended Gauss's work for the higher-order terms in order to improve extra-axial definition in telescopic photography. In 1873 Abbe added the sine-condition. Opticians who actually made the lenses chose the appropriate solutions presented in simplified forms by the mathematicians.

Lens-making

If the production of small objective lenses and mirrors became a routine and mechanized process in the nineteenth century, the grinding and polishing of large lenses and mirrors remained an art. Of course, as the diameters of objective lenses increased it became impossible to work them by hand. The large objective lenses made in the last half of the nineteenth century were, therefore, ground and polished on machines such as had been developed earlier for the grinding of mirrors. The glass blank was usually positioned on the lap which was mounted on a turntable driven by a small steam engine. The tool could be moved in the preferred pattern either by hand or by an eccentric arm driven by the steam engine.

It was in the fine corrections that the skill of the craftsman came to the fore. Once the lens had received its figure, it had to be tested. This had always been done by examining with it the image of a real or artificial star. In 1858, however, J. B. L. Foucault introduced the knife-edge test for examining the curvatures of lenses and mirrors. This test, named after him, and variations of it, not only allowed a very precise determination of the focal length of the lens, but also showed the smallest defects in the curvature and their exact locations. So minute were the defects thus uncovered that Howard Grubb checked whether the areas were too high or too low by making them expand or contract thermally. Passing his hand over the area five or six times or passing a camel-hair brush soaked in ether over it, sufficed to make the spot expand or contract sufficiently to indicate the exact direction and extent of the error.

But the corrections of these errors demanded the highest skills and great patience. After even the slightest local polishing the lens had to be allowed to reach thermal equilibrium and assume its true shape. This might take weeks. Alvan Clark would use nothing but his thumb for rubbing out minute local errors, insisting that no cloth or chamois was soft enough for this task. Visitors to Clark's shop reported that Clark often split his thumb open during this long and tedious process. The results, however, justified the efforts: the large objectives of this period had very nearly perfect curvatures.

Mountings and drives

As standards of accuracy became ever higher and the sizes and magnifications of refractors increased as well, mountings became more and more critical. A telescope weighing tonnes had to be kept from flexing; it had to be easily manoeuvrable in order to bring it to bear on the appropriate section of the sky; and it had to follow a celestial object automatically so that the observer had both hands free for making measurements. Moreover, minute vibrations transmitted through the mounting, a terrible nuisance even at low magnifications, became intolerable at high magnifications, virtually ruling out, for example, micrometrical work on double stars, planetary diameters, etc. These requirements taxed the skills of astronomers and engineers.

Whereas before 1840 the typical large refractor was still supported on the floor of the observatory, by 1900 all large instruments were supported by piers sunk into the bed-rock and independent of the observing floor. Even with these arrangements a telescope might nevertheless register the vibrations of the (often increasingly heavy) vehicular traffic in the vicinity. In the 1860s, for example, the Dudley Observatory in Albany had to be moved because of the frequent passing of heavy trains on the New York Central line at the bottom of the observatory's hill.

The arrangement of the telescope on the piers posed a great problem. Increasing weight placed greater and greater demands on the bearings, and engineers brought the lessons of the Industrial Revolution into the observatory. They designed all-metal tubes in which flexing was minimized and improved the systems of counter-weights, so that even instruments of record size could be moved at the touch of a hand. The form of mounting best suited for these large refractors was the equatorial mounting.

The idea of making one of the axes of rotation of an observational instrument parallel to the Earth's axis went back to Tycho Brahe, before the invention of the telescope. As mechanical drives became more desirable and even essential, especially after the introduction of photography, this form of

mounting became the preferred one. The two basic forms of equatorial mountings were the 'English' form in which the polar axis was supported at top and bottom, and the 'German' form, made popular by Fraunhofer, in which the polar axis was supported only at the bottom. With the former, stars near the pole could not be observed, while with the latter stars often could not be tracked through the meridian. These disadvantages, however, were more than offset by the great advantage of the automatic turning of the polar axis by means of a drive.

A particular form of the English equatorial – the fork mounting – came into use during this period. In this mounting the polar axis ends in an open fork in which the telescope tube is mounted on the declination axis. Only in reflecting telescopes, with the heavy mirror at the bottom of the tube, was this form of mounting practical. It was used by Foucault and W. Lassell in their reflectors, but the fork mounting was made popular by the 60-inch Mt Wilson reflector early in the twentieth century (see Chapter 8).

The simple clock drive regulated by a pendulum, popular early in the nineteenth century, gave a discontinuous motion, annoying in visual observations and unacceptable for photographic purposes. Fraunhofer and Foucault adapted the steam engine governor to the telescope. Fraunhofer put a hollow brass cone over the balls of the governor, and the balls rubbed against this with increasing friction as the speed of rotation increased. In Foucault's designs the rising balls caused friction to be exerted on the central shaft supporting them by means of linkages. In each case the motive force was supplied by a hanging weight.

The largest refractors

The 15-inch (38-cm) equatorial refractor installed in the newly founded Pulkovo Observatory in 1839 remained the largest refractor in the world until 1847, when it was joined by its twin, the Harvard refractor, also made by Merz & Mahler. Not until the 1860s was this aperture effectively exceeded, but once this happened, no telescope in the rest of the century was to claim leadership for more than a few years. The first refractor to surpass the Pulkovo and Harvard instruments was made not by a European, but by an American.

Alvan Clark (1804–87), a portrait painter in

3.1. The 80-cm photographic/50-cm visual refractor built for the Potsdam Astrophysical Observatory at the turn of the century by Steinheil and Repsold. Note the typical 'German' mounting.

Massachusetts, taught himself to grind lenses in his spare time, and gradually changed professions to become a telescope maker. Unacknowledged in his own country, Clark studied double stars with his own excellent instruments and attracted attention abroad. W. R. Dawes in England bought several Clark objectives (still of modest apertures), one of which, of 8-inch (20-cm) diameter, was used from 1860 to 1869 by William Huggins for his epoch-making work in spectroscopy. A trip to England in 1859 established Clark's reputation at home, and in the following year he was approached by officials of the University of Mississippi with an ambitious proposal.

Clark was asked to make a refractor with an aperture of 19 inches (48 cm) for the university. Up to that point, Clark's firm had only made objectives of apertures up to about 20 cm, and he was therefore hesitant to accept the proposal. Having examined the 15-inch objective of the Harvard refractor, however, Clark recognized its imperfections and thought he could do better. Accordingly he offered to make a 15-inch objective, as large as Harvard's, with the guarantee that, should it prove inferior to Harvard's, he would donate it to the University of Mississippi free of cost. The University's officials were, however, adamant. They wanted the largest refractor in the world, and, in the end, Clark agreed to make an $18\frac{1}{2}$-inch (47-cm) objective. This would represent a gain of about 50% in light-gathering power over the Harvard refractor.

In order to accommodate the work on this lens, Clark moved to larger facilities near Harvard. He ordered the crown and flint discs from the Chance Brothers firm, who, as he knew, had exhibited $18\frac{1}{2}$-inch discs of crown and flint of excellent optical qualities at the Paris Exhibition of 1855. The discs arrived in 1861 and by the beginning of 1862 Clark had finished the grinding and polishing. When his son Alvan Graham Clark (1832–97) tested the 8.2-m focus objective on 31 January 1862, he discovered the companion of Sirius predicted by Bessel. Before the objective had been mounted in its final form, therefore, its excellence had already been certified by a discovery.

By 1862, however, the American Civil War had broken out, and the University of Mississippi was unable to fulfil its agreement with Clark. Such a large and already celebrated objective could not lie idle for long. The Harvard Observatory competed with the Chicago Astronomical Society for it, and the Chicago group was able to raise the necessary funds by public subscription. Clark built a German equatorial mounting for the objective, and the finished telescope was installed in the Dearborn Observatory of the University of Chicago in 1866. There, G.W. Hough and S.W. Burnham used the instrument for their excellent work on Jupiter and double stars. In 1889 the refractor was moved to Northwestern University in Evanston.

The $18\frac{1}{2}$-inch Dearborn refractor did not for long remain the largest refractor in the world. Even as it was awaiting its installation, R.S. Newall, a wealthy English amateur, was making arrangements to have a larger instrument made. In 1863 he ordered crown and flint discs for a 25-inch (64-cm) objective from Chance Brothers. He persuaded Thomas Cooke (1807–68) in York to make the instrument for him. At this time Cooke was considered the finest manufacturer in England of refractors of up to 10-inch (25-cm) aperture. But the move from 10 to 25 inches proved to be very difficult for him. He had to develop new techniques to deal with the problems of size and weight in polishing the lenses. Thus, he floated the lenses on mercury during polishing, in order to avoid flexing. The finished objective had a focal length of 9.1m. The two lenses had a combined weight of 66 kg, and Cooke had to design a mounting that would support this weight with a minimum of flexing. He chose a German equatorial form to

support the tube and devised his own drive in which a conical pendulum served as regulator.

Altogether, the Newall refractor took seven years to complete, and it broke Cooke's health. In 1871, three years after Cooke's death, the telescope was finally erected on Newall's estate at Gateshead-on-Tyne. Newall's original plan to have the instrument moved to a better climate after initial testing at Gateshead was frustrated for almost twenty years. Finally, in 1890, the year following Newall's death, the 'Newall refractor' was transferred (at considerable cost) to the Solar Physics Observatory at Cambridge, where it saw much more intensive use under somewhat better climatic conditions. In 1956 it was presented to the National Observatory of Athens, where it is still in operation.

By 1871, however, work on an even larger refractor was already under way. The US Naval Observatory at Washington had from its beginning specialized in position measurements because its function had been rigidly restricted to practical matters of navigation by Congresses unwilling to spend public money on research without a clearly foreseen practical benefit. In the post-Civil-War Republican Congress, however, the climate was more sympathetic to basic research, and in 1868 Simon Newcomb began a campaign for a large refractor that resulted in 1870 in an appropriation of $50000 for this purpose.

Alvan Clark had refigured some of the objectives of the smaller telescopes at the Naval Observatory with impressive results, and it was to him, therefore, that Newcomb now turned. Clark had been eager to make an instrument that surpassed the Newall refractor, and he had just signed a contract with Leander McCormick to make the largest refractor in the world for the University of Virginia. McCormick graciously allowed the Naval Observatory priority, and Clark now agreed to make 26-inch (66-cm) objectives for both observatories. After several failures, Chance Brothers delivered the crown and flint blanks late in 1871, and by the summer of 1873 the lenses were finished. The crown component was less than 5 cm thick at its centre, and the flint component less than 2.5 cm. Their combined weight was less than 50 kg, considerably less than Cooke's objective for the Newall refractor. Nevertheless, Clark had to abandon the use of wood for the tube. He had

3.2. R.S. Newall's 25-inch refractor, built by Thomas Cooke & Sons around 1869, was for a few years the largest in the world. Newall erected the telescope on his estate, a site so unfavourable that during fifteen years he had only one night in which he could advantageously use the full aperture.

inspected the Newall refractor, and he now built a German equatorial mounting modelled after the Newall's mounting to accommodate the 26-inch, 9.9-m focus objective. By late 1873 the new instrument was installed in the Naval Observatory. It lived up to expectations: in 1877 Asaph Hall discovered Phobos and Deimos with the Naval Observatory refractor. Its identical twin, built for McCormick, went into operation at the University of Virginia in Charlottesville in 1884.

In Vienna, the astronomer K. L. v. Littrow, director of the old observatory, received authorization in 1872 to plan a new observatory in the Vienna suburbs. Littrow wanted a very large refractor, and he awarded the contract to Howard Grubb (1844–1931) of Dublin. Up to this point, Grubb and his father Thomas Grubb (1800–78) had specialized in reflectors, and one result of this was that they were experienced in mounting heavy instruments. Howard Grubb designed a German equatorial mounting that would allow the astronomer to follow a star through the zenith. The many innovations and especially the sturdiness of this design made it a great improvement over the mountings supplied by the Clark firm, and it became a model for all future mountings of large refractors.

Grubb's agreement with Littrow, made in 1875, called for a 27-inch (69-cm) objective. Chance Brothers refused to attempt making the requisite discs for such an objective, and Grubb turned to Feil in Paris. Not surprisingly Feil encountered some difficulties, and only in 1879 did Grubb receive the crown and flint discs. By this time he had already finished the mounting and the ancillary equipment. The objective was now ground and polished, and the telescope went into operation in the new Vienna Observatory in 1881.

In the meantime, Otto W. Struve, director of the Pulkovo Observatory, had been authorized by the Russian government to acquire a very large refractor for his observatory. Struve contracted with the Repsold firm in Hanover for the building of the telescope, but could not reach agreement with any European firm on the objective. He therefore travelled to America, where Newcomb showed him the Naval Observatory refractor and warmly recommended the Clark firm. In 1879 Struve ordered a 30-inch (76-cm) objective from Clark. Feil again had trouble making the appropriate

discs, especially the flint disc, and the objective was not completed until 1884, by which time Repsold had finished the mounting. The world's largest refractor (with a focal length of 14.1 m) went into operation at Pulkovo in 1885. It was used in studies of double stars and the proper motion of Procyon.

The Pulkovo refractor did not retain its lead for long. In San Francisco, the heirless businessman James Lick had decided to immortalize himself by endowing the largest telescope in the world. Scientific advisors counselled him to choose a refractor, and designs and estimates were gathered. But Lick's strong-mindedness and his ignorance of the needs of astronomers delayed action until after his death in 1876. Mt Hamilton near San Francisco was now chosen as the site of an observatory, and the observatory was designed by Simon Newcomb and E. S. Holden, who was later to become its director.

In 1880 the Clark firm was given a contract to produce a 36-inch (91-cm) objective and a photographic corrector. Clark ordered the appropriate blanks from Feil in Paris, but that firm had great difficulty producing them. The flint blank caused few problems: it arrived in Massachusetts quickly, and the Clarks had the flint lens finished in 1882. The crown blank, however, required no fewer than nineteen tries and took $3\frac{1}{2}$ years to produce; the Clarks finished the grinding and polishing of this component in 1885. The disc for the 33-inch (84-cm) photographic corrector was rejected by the Clarks when they detected imperfections in it. When Feil guaranteed the disc, they attempted to shape it, but it broke, and a new disc was supplied free of cost. The corrector was in consequence finished only in 1887.

The equatorial mounting was designed and built by the Cleveland engineering firm of Warner & Swasey. Warner & Swasey were relative newcomers in this endeavour, but the firm had established itself as one of the finest machine-tool makers in the world, and it was equal to the task. The German equatorial mounting was erected at Mt Hamilton in 1887 (on a pier containing Lick's body), and the great refractor began operations in 1888. Its focal length was 17.6 m, the separation between the crown and flint components in their cast-iron cell was 15 cm, and the 84-cm photographic corrector had a focal length of 14.6 m.

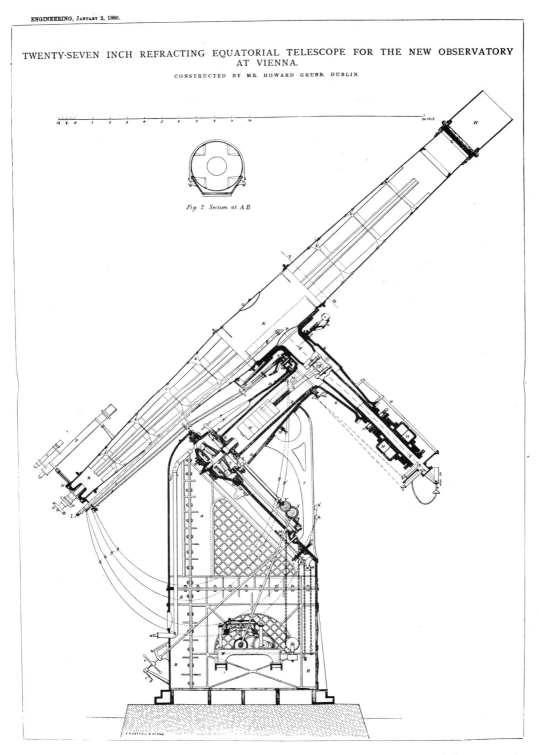

ENGINEERING, January 2, 1880.

TWENTY-SEVEN INCH REFRACTING EQUATORIAL TELESCOPE FOR THE NEW OBSERVATORY AT VIENNA.

CONSTRUCTED BY MR. HOWARD GRUBB, DUBLIN.

Fig. 2. Section at A.B.

3.3. In 1881 this 27-inch Grubb refractor in Vienna surpassed the 26-inch Clark at the US Naval Observatory, becoming the model of subsequent mountings for large refractors.

3.4. Pulkovo 30-inch, 1885, unsurpassed until the Lick 36-inch refractor in 1888.

Because of an excellent support system, the fine viewing conditions of Mt Hamilton and the exceptional staff of the observatory, the Lick refractor quickly established itself as one of the most productive instruments in the history of astronomy.

With the completion of the Lick refractor, representatives of the University of Southern California at Los Angeles asked the Clark firm to make an objective for a telescope larger than the Lick refractor for a projected observatory on Mt Wilson near Los Angeles. In 1889 the Clarks agreed to make a 40-inch (102-cm) objective, and the blanks were ordered from Mantois in Paris. Mantois had few difficulties in making the discs, but when these arrived at the Clark works in Cambridge, Massachusetts, the University of Southern California was unable to pay for them. The opportunity to obtain the largest refractor in the world appealed to many other institutions, but the necessary sum for the objective, mounting and building – at least $200000 – was not easy to raise on short notice. The race was won by the University of Chicago.

George Ellery Hale had just been appointed Associate Professor of Astral Physics and Director of the Observatory at the University of Chicago in 1892, when he managed to convince the streetcar magnate C.T. Yerkes to donate the money for the telescope. Alvan Graham Clark, the last surviving member of the family, now began figuring the lenses, and Warner & Swasey were asked to supply the mounting. The German equatorial mounting was finished in 1893 and was displayed at the Columbia Exhibition in Chicago that year. Williams Bay, Wisconsin, 150 km from Chicago was chosen as the site for the new observatory, and the foundation for the Yerkes Observatory was laid in 1895. The Observatory with its 40-inch refractor (focal length 19.3-m) began operation in 1897. The Yerkes refractor was the largest in the world and remains so today.

Had the limit of refractors been reached? When J.E. Keeler had examined the Yerkes objective at the Clark works as an 'expert agent' in 1896, he wrote in his report (later published in *The Astrophysical Journal*):

From these tests it appears that the character of the image varies with position of the lenses relative to each other, and, to a less extent, with the position of the objective as a whole relatively to its cell. It is probable that flexure of the lenses is the principal cause of the observed changes, and it is interesting to note that there is here evidence, for the first time, that we are approaching the limit of size in the construction of great objectives.

Although no such effect was actually apparent once the lenses had been mounted permanently, weight was beginning to pose a problem: the combined weight of the two components of the Yerkes objective was over 225 kg, though the engineering skills of Warner & Swasey were equal to the challenge. In all respects, the Yerkes refractor lived up to expectations.

Before his death in 1897 Alvan Graham Clark had declared his aim to make a 60-inch (152-cm) objective. Others shared his optimism, and in the twentieth century several larger refractors were, in fact, attempted. The reason why, after Yerkes, large refractors fell out of favour lay in the changing nature of astronomy itself. As we shall see, the needs of spectroscopy and photography for large telescopes were better served by reflectors than refractors, and by the turn of the twentieth century reflector technology had advanced sufficiently to allow reflectors to take the lead in this field.

Meanwhile the experience gained in attempting very large refractors was applied to more modest instruments, some of which were specifically designed for photography. Whereas in 1850 the largest apertures were 38 cm, by 1880 such apertures were becoming common in observatories, so much so that by the end of the century there were well over 60 refractors with apertures in excess of 35 cm. This is not to say, however, that making objectives of such sizes had become common skill: rather, a handful of firms (the Clarks in Massachusetts, Grubb in Dublin, the Henry brothers in Paris and the Merz firm in Munich) accounted for the vast majority of these refractors. In addition, mountings had become increasingly sophisticated. Whereas earlier the opticians had usually provided their own mountings for refractors, towards the end of the century firms such as Gautier in Paris (working with the Henrys) and Warner & Swasey in Cleveland (working with the Clarks and later with John A. Brashear) had become specialists in this field. Although Sir Howard Grubb and the Merz firm continued to

3.5. The mounting for the 40-inch Yerkes refractor on display at the Columbia Exposition in Chicago in 1893.

supply their own mountings, the building of large refractors had, for the most part, become a cooperative venture by the turn of the century.

The first efforts towards standardizing refractors came in connection with the international cooperative project of the *Carte du Ciel*, which the preceding chapter has recounted in detail. At the organizing conference in Paris in April 1887, the participants agreed to use as a prototype the photographic refractor of 34-cm aperture made by the Henry brothers and mounted by P. Gautier at the Paris Observatory. By 1895 seventeen almost identical photographic refractors were in operation in locations all over the world. Nine of these were made by the Henrys and Gautier, six by Sir Howard Grubb and two jointly by Steinheil in Munich and Repsold in Hamburg.

Encouraged by their success, the Henrys and Gautier went on to build the 76-cm visual refractor of the Nice Observatory (focal length 18.0 m) in 1886, and the great photographic–visual double for the Meudon Observatory in 1896. This last instrument, following the design of the *Carte du Ciel* astrographs, had an 83-cm photographic refractor with a focal length of 16.2 m and a 62-cm visual guide scope with a focal length of 15.9 m. It has remained the largest photographic refractor in the world. A few years later Steinheil and Repsold collaborated on the 80-cm-photographic, 50-cm-visual refractor for the Potsdam Astrophysical Observatory. The crown and flint discs for this instrument were provided by Schott in Jena. This instrument, which went into operation in 1901, was also designed for spectrographic work.

The design of photographic objectives presented many problems to opticians. Sir Howard Grubb, the Henrys, Steinheil and others had to investigate entirely new lens designs, and much of this design involved long trial-and-error procedure. By the middle 1890s, however, these design problems were solved, and astrographs now became more common. Grubb's first large astrograph was the 26-inch (66-cm) instrument for the Greenwich Observatory. In 1889 E. C. Pickering, director of the Harvard College Observatory, secured a gift of $50000 from Miss C. W. Bruce for a photographic refractor to survey the southern skies. The 24-inch (61-cm) Bruce astrograph, finished by the Clark firm in 1893, had the low aperture ratio of f/5.5, and a much larger field of view than the *Carte du Ciel* astrographs. Upon its installation, in 1896, at Harvard's southern station at Arequipa, Peru, this instrument enabled the Harvard staff to photograph the entire southern sky relatively quickly. It was a model for later short-focus astrographs.

Reflectors

The revelation of the spiral structure of nebulae in 1845 by the great reflector of the Earl of Rosse was further proof of the potential of large reflectors with their immense light-gathering power. There were, however, good reasons why astronomers interested in serious, long-term observing projects would reject reflectors for precision observatory work. The tarnishing of the speculum-metal mirror meant that periodically an observing project had to be interrupted in order to repolish the mirror. This entailed refiguring its curvature, and it was commonplace that large reflectors performed up to expectations only in the hands of their original makers, who could repolish them successfully. Moreover, in the middle of the nineteenth century supporting the great weight of the mirror still presented major problems. Flexure and distortion of the mirror, due to inadequate support and variations in temperature, meant that images were distorted. No convenient and accurate mounting that could cope with these problems yet existed, and thus the reflector was an awkward instrument, useless for precision work in which observatories specialized. There was the added disadvantage that the tube was open to the air, so that air currents in the tube interfered with seeing, while the mirror often collected dew. Before the reflector could take its place in the observatory, these problems had to be solved.

By the middle of the century, the engineering profession could, however, begin to deal with the problem of mounting heavy mirrors. Significantly, it was the Manchester engineer J. Nasmyth (1808–90) who first mounted a substantial mirror conveniently. In 1842 he built a 20-inch (51-cm) Cassegrain–Newtonian reflector in which the tube was mounted on trunnions, much like a cannon. A 45° plane mirror reflected the light through one hollow trunnion to the eyepiece. The altazimuth mounting was on a turntable which the observer could turn by moving a wheel while he remained seated at the eyepiece.

If Nasmyth's design was idiosyncratic (although

3.6. The Bruce 24-inch doublet astrograph built in 1893 by Alvan Graham Clark for photographic reconnaissance at Harvard College Observatory.

convenient), his Liverpool friend W. Lassell (1799–1880) profited from Nasmyth's engineering skills in designing the first equatorial mounting for a large reflector. Lassell began by providing a 9-inch (23-cm) Newtonian with an equatorial fork mounting in the early 1840s, and then applied the same type of mounting to a 24-inch (61-cm) Newtonian. With this last instrument he discovered four additional satellites of Uranus and Neptune. In 1852 he transported the instrument to Malta where the atmospheric conditions allowed

the large aperture to be used to greater advantage, and in 1861 Lassell erected there a 48-inch (122-cm) Newtonian, again fork mounted. This instrument indicated the feasibility of mounting large reflectors equatorially, and Lassell's discoveries were seen as proof of the important contributions to astronomy to be expected from large-aperture reflectors. In his three reflectors Lassell also addressed himself to the problem of flexure of the mirror at various positions. He provided the 9-inch mirror with a single pad support attached to a lever

3.7. The 48-inch Newtonian reflector built by William Lassell and used in Malta 1861–4.

with a heavy weight at its end. When the telescope was in the horizontal position the lever exerted no force on the pad; when the tube was in the vertical position the lever exerted a force equal to the weight of the mirror. In his larger telescopes he used two systems of levers, one for support in the horizontal and one for support in the vertical position. After Lassell such systems of levers became an accepted feature in reflectors; they went far towards solving the problem of mirror flexure with position.

The last great reflector with a speculum-metal mirror was the Melbourne Reflector. For some time, British astronomers had felt the need for a large telescope in the southern hemisphere to study southern nebulae. The Royal Society had

taken the lead in this project, but it had been unable to secure the necessary funds from Parliament. In 1862 the authorities of the colony of Victoria turned to the Royal Society for advice concerning a large telescope for the new observatory at Melbourne. A committee consisting of Lord Rosse, W. De la Rue and T. R. Robinson recommended a large reflector, and in 1865 the legislature of Victoria voted the appropriate funds. Thomas Grubb was given the contract, and the committee oversaw the work.

For the study of nebulae a large aperture was needed, and the committee decided on a 48-inch (122-cm) reflector. This was the largest mirror deemed practical for an equatorial mounting. A refractor with comparable light-gathering power would need an aperture of about 90 cm, and such an instrument was, at that time, out of the question financially and improbable practically. The committee felt that the silver-on-glass process (see below) had not been sufficiently proven to warrant the risk in this important project. The mirror was, therefore, to be made of speculum metal. For ease of observation the Cassegrain type was preferred to the Newtonian type.

Grubb experienced only one failure in casting the two specula. He built a special machine for polishing them, and their shapes were judged perfect when they were finished in 1868. Their surfaces were then varnished to protect them during the long voyage. Grubb chose an English equatorial mounting, minimizing the load on the lower bearing by means of a system of counterweights, so that the instrument could be manipulated with great ease. By the end of 1869 the reflector had been erected in Melbourne (see Figure 9.4).

Problems immediately beset the astronomers there. As early as January 1870 their report to the Royal Society indicated that upon the removal of the varnish the surfaces of the mirrors had begun to tarnish rapidly "with evident effect on the performance". Repolishing was a very risky undertaking in Melbourne, where the necessary expertise was lacking. R.L.J. Ellery taught himself the art of polishing mirrors as best as he could, but his attempts at repolishing the great mirrors were not entirely successful. Although some useful work was done with the Melbourne Reflector, it did not live up to its expectations, and indeed, it signalled the end of the great speculum-metal reflectors.

In the meantime, a new type of mirror was being developed. As early as 1827 G.B. Airy had suggested the use of silver-coated glass mirrors in telescopes, but it was not until the 1850s that the state of the art of chemistry could accomplish the coating process. In 1856 C.A.v. Steinheil and J.B.L. Foucault independently produced small paraboloidal glass mirrors on which a thin coat of silver had been deposited by a precipitation method found by J.v. Liebig.

This new type of mirror presented practitioners with several difficult problems, however. Although the optical quality of the glass was not of paramount importance for these mirrors (since only the silver coating was part of the optical system), the unidirectional structural property of plate glass, caused by its method of manufacture, could cause astigmatism. The glass disc had to be parabolized accurately, in order to eliminate spherical aberration, and this proved to be much more difficult for glass than for speculum metal in larger mirrors. Finally, a way had to be found to give the silver film the proper consistency, so that it could be brought to a high gloss with gentle polishing and would not tarnish too rapidly. These difficulties were overcome slowly by the combined efforts of a number of practitioners over the next thirty years.

Before his death in 1868, Foucault himself made a number of successful mirrors, using his own knife-edge test to check the curvature of the glass. His largest was an 80-cm mirror. In England G.H. With (1827–1904) was successful in parabolizing mirrors of up to 46 cm in diameter. He combined efforts with the instrument-maker J. Browning (1835–1929), well known for his excellent spectroscopes, to market a range of silver-on-glass Newtonian reflectors with apertures up to 15 inches (38 cm), mounted as altazimuths or equatorials. Browning became an enthusiastic advocate of the reflector: his little book *A Plea For Reflectors*, first published in 1867, went through a number of printings. Browning and With made excellent reflectors of modest apertures and brought these instruments within the financial grasp of many who could not afford refractors of comparable apertures.

The reflector's freedom from chromatic aberration and its shorter focal length became great advantages with the development of photography and spectroscopy. In New York, Henry Draper (1837–82) began by grinding speculum-metal mirrors in the late 1850s, but he quickly turned to silver-coated glass mirrors. By 1862 he had finished a $15\frac{1}{2}$-inch (39-cm) reflector, and besides taking many pioneering photographs with this instrument, he wrote a number of papers on how to make these new reflectors. In 1866 he began on a 28-inch (71-cm) reflector which after much grinding and polishing he finally finished in 1872. His efforts were quickly rewarded with a plate of the spectrum of Vega, the first stellar spectrum to be photographed.

By the late 1870s many of the problems of constructing silver-on-glass reflectors had been solved. The best estimates indicated that beyond about a metre in aperture, mirrors would have the advantage over lenses because of the increasing absorption in lenses through the greater thickness of glass. What was needed now was a clear demonstration that very large silver-on-glass reflectors with none of the disadvantages of the Earl of Rosse's great reflector could, indeed, be made. The efforts of A. A. Common (1841–1903) went a long way towards that goal.

Common, a wealthy engineer, began his hobby of astronomical photography with a small refractor in the early 1870s. When he wanted an instrument of larger aperture, he chose the reflector, and by 1876 G. Calver (1834–1927) had made for him an 18-inch (46-cm) silver-on-glass mirror, which Common mounted equatorially as a Newtonian. He was well pleased with this instrument and became a vocal advocate of the new form of reflector. Calver now started on a 36-inch (91-cm) mirror for Common. The grinding and polishing proved to be very troublesome, and Calver tried no fewer than five machines in this task. This done, Common silvered the mirror himself and erected it as a Newtonian on a fork-type equatorial mounting in his garden in Ealing near London. The mounting had several innovations, chief among which was the partial floating of the polar axis in mercury to reduce friction. This instrument, finished in 1879, was a great success, and Common proceeded to take a series of important photographs with it, culminating in 1883 in several exposures of the Orion Nebula showing stars not seen by visual observation (see Figure 2.5). For these photographs he received the Gold Medal of the Royal Astronomical Society in 1884.

Calver went on to make a number of large mirrors, the largest of which, a 50-inch (127-cm) effort for Sir Henry Bessemer, met with failure as a result of Bessemer's interference. For his part, Common made his own mirrors, starting in 1886. In 1880 he had ordered, from the St Gobain glass works in France, a 61-inch (155-cm) glass disc with a central hole for a Cassegrain. The disc arrived in 1882, but not until after finishing his series of photographs with his 36-inch instrument did Common begin work on this new mirror. In 1885 he sold his 36-inch to E. Crossley (1841–1905) of Halifax, Yorkshire, to make room for the 5-foot. Having no experience in glass-working, Common learned the craft by constantly regrinding and

polishing the 5-ft disc. The instrument was finished in 1889, but Common was unhappy with the images, and he ordered a new disc, without a hole this time, which he figured in three months in 1891. In its new form the instrument was used as a Newtonian. The polar axis was a hollow wrought-iron cylinder which floated partially in a tank of water, and the telescope was supported by a fork arrangement, as in the 36-inch instrument. During this period Common also produced several large mirrors (up to 91 cm) for other astronomers.

Although Common's 5-ft reflector was excellent, it produced few useful results. Common's attention was diverted to other engineering projects shortly after finishing the instrument, and it languished in his garden, so that not for some time was the excellence of his reflectors demonstrated. In 1893 Crossley retired from astronomy and shortly afterwards gave Common's 36-inch reflector to the Lick Observatory, where, he hoped, its aperture could be used to greater advantage than in the poor climate of Yorkshire. Installed on Mt Hamilton in 1896, the 'Crossley reflector' became the centre of a controversy between Holden and certain members of his staff who thought the instrument was too clumsy for serious work. When Keeler assumed the directorship in 1898, he improved the Crossley reflector and initiated a programme of nebular photography with it. After Keeler's death, C. D. Perrine continued the project, completely remounting the instrument between 1902 and 1905. The resulting photographs helped establish the reflector as a serious observatory instrument.

Common's 5-ft reflector was bought by the Harvard College Observatory in 1904 for $20000. Pickering and his staff laboured for ten years trying to improve the mounting before they abandoned the effort in 1914. Harlow Shapley had no more success in 1922. One of the mirrors finally was installed in an entirely new mounting at Harvard's new Boyden Station near Bloemfontein, South Africa, in 1927, where it is still in service. The other mirror served briefly, from 1933 to 1937, as the primary receptor of a new reflector built for Harvard's Oak Ridge Station in Massachusetts, before being replaced by a pyrex mirror.

Although for many purposes the refractor was still the preferred instrument, the reflector was coming into its own at the end of the nineteenth century. Its greater aperture was essential for the

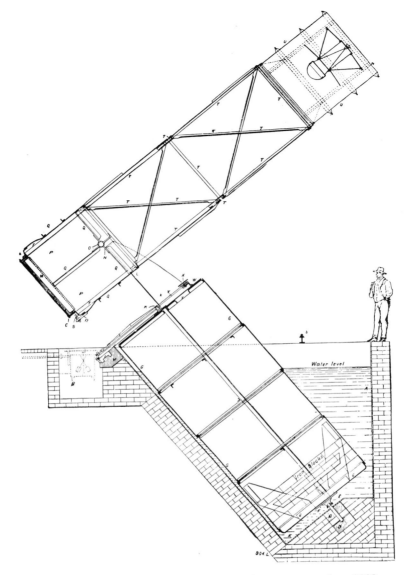

3.8. A.A. Common's 5-ft reflector with its flotation mounting, 1889.

study of nebulae, which was now becoming a crucial area in astronomy, and its lack of chromatic aberration and short focal length were essential characteristics for the now rapidly developing science of astrophysics.

Further reading

Brief, often illustrated, articles on individual telescopes can be found in the contemporary technical literature, usually within a year of the dedication of the instrument. The following periodicals are especially useful in this regard: *The Astrophysical*

Journal, Engineering, Monthly Notices of the Royal Astronomical Society, Nature, The Observatory, Scientific American, and *Zeitschrift für Instrumentenkunde*. Note the list of large telescopes in the Appendix to this volume.

André Danjon and André Couder, *Lunettes et Télescopes* (Paris, 1935)

Henry C. King, *The History of the Telescope* (London and Cambridge, Mass., 1955)

J. A. Repsold, *Zur Geschichte der astronomischen Messwerkzeuge von 1830 bis um 1900*, vol. 2 (Leipzig, 1914).

Rolf Riekher, *Fernrohre und ihre Meister* (Berlin, DDR, 1957)

Deborah Jean Warner, *Alvan Clark & Sons: Artists in Optics* (Washington, DC, 1968)

4

The new astronomy

A.J. MEADOWS

In 1888, a popular science book entitled *The New Astronomy* appeared in the United States. It aroused considerable interest and was read widely, and not only in the USA. Its author, S.P. Langley (1834–1906), explained that his book had a particular focus of astronomical interest:

> . . . *within a comparatively few years a new branch of astronomy has arisen, which studies Sun, Moon, and stars for what they are in themselves, and in relation to ourselves. Its study of the Sun, beginning with its external features (and full of novelty and interest, even, as regards those), led to the further inquiry as to what it was made of, and then to finding the unexpected relations which it bore to the Earth and our own daily lives on it, the conclusion being that, in a physical sense, it made us and re-creates us, as it were, daily, and the knowledge of the intimate ties which unite man with it brings results of the most practical and important kind, which a generation ago were unguessed at.*
>
> *This new branch of inquiry is sometimes called Celestial Physics, sometimes Solar Physics, and is sometimes more rarely referred to as the New Astronomy.*

Langley was thus setting out to popularize what would nowadays be called 'astrophysics'. The title *The New Astronomy* derives from Kepler. Langley's use of it indicates that he saw nineteenth-century astrophysics as a breakthrough equivalent to seventeenth-century work on planetary motions. His emphasis, however, is clearly different from ours today: over a hundred pages of the book are devoted to the Sun, rather less to the remainder of the solar system, and only thirty pages to objects outside the solar system. This weighting partly reflects Langley's own interests, but it is undoubtedly true that solar studies initiated and guided much of astrophysics in the early years of the subject. The main reason, of course, was the brightness of the Sun: early astrophysical instruments were not very efficient, and so could not readily be used with feeble light sources. A subsidiary reason was that the Sun presented a large disc, capable of being studied in detail. (This was also true of the Moon: but the Moon changed little, if at all, whereas the Sun was perpetually producing new phenomena to observe.) Additionally, as Langley noted, solar–terrestrial interaction had been known since the mid-nineteenth century, and was being eagerly investigated. Finally, a proper understanding of the Sun was expected to illuminate the physical properties of all stars.

These nineteenth-century arguments for the priority of solar work in early astrophysics are obviously valid. New branches of astrophysics in the twentieth century – such as ultraviolet and X-ray – have tended to concentrate initially on the Sun for the same reasons. In looking at the growth of astrophysics in the last decades of the nineteenth century, we shall therefore be concentrating on solar physics , and, at the same time, on the most fruitful of the new techniques introduced to study the Sun – spectroscopy.

Astrophysics and other disciplines

It is symbolic that the most obvious starting point for the origin of astrophysics – the work of Kirchhoff and Bunsen – should have been the result of cooperation between a physicist and a chemist. Astrophysics could only develop when the physical sciences, of which it is part, had reached a sufficient stage of maturity in both their theoretical and experimental techniques to assist its progress.

An example of this interaction with physics and chemistry is provided by the discovery of the series of hydrogen lines in the visible region of stellar

4.1. Sunspot of 5 March 1873 drawn by S.P. Langley.

spectra. The first few hydrogen lines were well known, both astronomically and in the laboratory, by the 1870s, but it could hardly have been demonstrated conclusively that they formed a regular sequence. Then, in 1880, William Huggins announced that he had found additional lines in his photographs of stellar spectra.

In the spectra of such stars as Sirius and Vega there came out in the ultra-violet region, which up to that time had remained unexplored, the completion of a grand rhythmical group of strong dark lines, of which the well known hydrogen lines in the visible region form the lower members. Terrestrial chemistry became enriched with a more complete knowledge of the spectrum of hydrogen from the stars. Shortly afterwards, Cornu succeeded in photographing a similar spectrum in his laboratory from earthly hydrogen.[1]

The hydrogen lines were now so obviously seen to be following a regular pattern that the search for some kind of formula to explain their relative positions intensified. In 1885, J. J. Balmer proposed a formula that, despite its simplicity, predicted the wavelengths of the lines with good accuracy. This pinning down of the 'Balmer' series of hydrogen represented the first important breakthrough in the systematization of line spectra.

Langley classified the major areas of the new astronomy as photometric, spectroscopic and photographic. In all of these, major advances in instrumentation often had little to do with astronomy. In fact, some can be related directly to industrial or social demands. Thus, developments in both photometric and spectroscopic instrumentation were stimulated from the mid-nineteenth century onwards by the growth of artificial lighting. The need to measure, and to enhance the efficiency of, illumination by gas and electricity led to much optical experimentation. At the same time, the interest in portraiture speeded the development of photographic emulsions. Perhaps the most striking example of this occurred in 1882, when David Gill at the Cape Observatory borrowed a camera from a local photographer to take pictures of a bright comet. He not only obtained good photographs of the comet, but also, unexpectedly, of the star background. In consequence, astronomers were stimulated to develop photographic star atlases and catalogues. Again there was a reciprocal interaction, in the sense that astronomers, while benefiting from work in other fields, themselves contributed to the progress in instrumentation. For example, J.F.K. Zöllner in Germany carried out important development work on both photometric and spectroscopic instrumentation.

In summary, the last thirty years of the nineteenth century saw instrumentation and techniques sufficiently developed for a variety of astrophysical investigations to gather momentum. As we shall see, the main retarding factor was the lack of an adequate theoretical basis for interpreting the observations. First, however, we must look at some of the communal and institutional problems facing the early astrophysicists.

Astrophysics and the astronomical community

If we examine the backgrounds of the men who pioneered the new astronomy, they typically came from professional families. But, in this, they differ less from the classical astronomers than in their education. In Britain, the education of an Astronomer Royal (and of his Chief Assistant, too) was typically based on the Mathematics Tripos at Cambridge. By way of contrast, none of the pioneer British astrophysicists underwent this training. Some of the greatest – for example, Huggins, Norman Lockyer and W. De la Rue – were essentially self-educated. A similar absence of a specialized mathematical education can be detected in early astrophysicists in other countries. Thus, in the USA, from the last quarter of the nineteenth century onwards, astrophysicists increasingly came from a university background in physics: a trend which has, of course, strengthened during the twentieth century.

While new recruits were being drawn into astronomy via astrophysics, some, though by no means all, of the practitioners of classical astronomy were less impressed. Their suspicions had more than one cause, and did not dissipate rapidly. Some astronomers felt that astrophysical methods were less direct than those of classical astronomy, and so less trustworthy. S. C. Chandler still insisted in the 1890s that, for inclusion in a catalogue of variable stars, any star whose variability had been

detected by astrophysical means, should be checked by visual observations. A stronger objection was that the new astronomy was less precise, and therefore less 'scientific', than classical astronomy. In 1886, Otto W. Struve wrote to the Academy of Sciences in St Petersburg:

> As yet, astrophysical investigations are far from the standard of scientific accuracy possessed by classical astronomy, which, with its solid mathematical base and constant progress in both observation and theory, rightfully occupies the premier place among the experimental sciences. God forbid that astronomy should be carried away by a fascination with novelty and diverge from this essential basis, which has been sanctified for centuries, and even millennia.[2]

The feeling that astrophysics was the more superficial branch of the subject persisted into the twentieth century. In writing to J. Trowbridge on 9 December 1914, Percival Lowell referred to "astrophysics which, of late years, owing to the effect that pictures have on people, has usurped to itself the lime-light to the exclusion of the deeper and more profound parts of astronomy proper". This attitude diminished after World War I as a result of both the dethronement of Newtonian theory and the growth of a coherent theoretical basis for astrophysics.

Finally, there was a feeling that resources, which should have been devoted to classical astronomy, were being diverted to the new astronomy. In particular, stellar astrophysics needed the largest telescopes possible, which, being mainly refractors, were also in demand by classical astronomers. In 1870, Asaph Hall noted in the *American Journal of Science* that:

> . . . in the case of Saturn, Uranus and Neptune it appears to me that the instrumental means are already at hand for making an accurate determination of their masses, and a more complete investigation of the theories of their satellites. When the novel and entertaining observations with the spectroscope have received their natural abatement and been assigned their proper place, it is to be hoped that some of the powerful telescopes recently constructed may be devoted to this class of observation.

Some classical astronomers at least had state-supported observatories in which to work, but how could a proto-astrophysicist afford to pursue his interest? A fair sprinkling of the pioneers had sufficient wealth both to purchase the necessary equipment, and to work full time at the subject, themselves. Such were De la Rue and Huggins in England and Henry Draper and L.M. Rutherfurd (and, later, George Ellery Hale) in the United States. But most astrophysicists had to turn elsewhere for funds. In his Preface to *The New Astronomy*, Langley complained:

> It is not generally understood that among us not only the support of the Government, but with scarcely an exception every new benefaction, is devoted to 'the Old' Astronomy, which is relatively munificently endowed already; while that which I have here called 'the New', so fruitful in results of interest and importance, struggles almost unaided.

Langley was to some extent indulging in special pleading here: in 1890 his arguments for the support of astrophysics in the USA were different.

> Briefly stated, the work for which the older government observatories at Greenwich, Paris, Berlin and Washington were founded, and in which they are actually chiefly engaged, is the determination of relative positions of heavenly bodies and our own place with reference to them. Within the past twenty years all these governments but our own have created an addition to these, a distinct and additional class – astrophysical observatories, as they are called.[3]

State finance had, in fact, been involved in the foundation of astrophysical observatories in Potsdam, Meudon and South Kensington (in that order) as early as the 1870s. Although state and federal aid may have been less forthcoming in the USA, the founding there of astrophysical observatories by private donation ultimately proved at least as fruitful, leading to the creation of a series of major observatories committed to astrophysics – Lick in 1876, Yerkes in 1897 and Mt Wilson in 1904. At the same time, in most countries the old-established government observatories sooner or later became involved in some astrophysical work. For example, E.W. Maunder started solar observations at Greenwich in 1873, and soon expanded his work into other areas of astrophysics. Finally, state aid was necessary for financing most of the eclipse expeditions which were a major feature of solar physics in the latter part of the nineteenth century. Thus early eclipse expeditions dispatched from Britain were organized by the Astronomer Royal, and government ships were requisitioned to transport the observers.

Astronomical spectroscopy

The earliest spectroscopes were based on the prism, and such instruments remained popular throughout the nineteenth century. However, they had obvious drawbacks; in particular, the dispersion was non-linear, so that each spectroscope needed to be calibrated in terms of wavelength. In addition, the prisms absorbed some of the incident radiation, especially in the ultraviolet. Both these problems could be overcome by using gratings, instead of prisms. A.J. Ångström had already emphasized the need to determine wavelengths from grating measurements in the 1860s, but the problem of how to manufacture large, high-quality gratings remained. The art of producing such gratings was developed especially in the United States; the first person to establish a major reputation in this field being L.M. Rutherfurd (1816–92). His gratings were highly sought after, and were obtained by any means possible, as the following letter of 3 May 1879 from Langley to Lockyer suggests.

> On reaching New York I called at Mr Rutherfurds to see about your grating. He is away, but Chapman told me he (Chapman) had sent you one, two weeks ago which I suppose you have now got. I seized upon two large ones I found there and bore them off, that being the only way of getting them, as he has apparently promised so many people that he has no longer any very exact idea of the order of priority of claimants for the few he makes.

Rutherfurd was followed by H.A. Rowland (1848–1901), who developed the grating to something like its modern capabilities, introducing the more efficient concave grating in the 1880s. In the following decade, he used some of these gratings to map the solar spectrum in detail between 3000 and 7000 Å. (The dispersion was sufficiently high for his final map to have a total length of over twelve metres.) Consequently, by the end of the nineteenth century, wavelength standards had been established which, with minor modifications, served astrophysics for many years. Part of the significance of this laborious exercise will become apparent later, when we discuss line coincidences.

4.2. Apparatus for comparing solar and laboratory spectra as depicted by Norman Lockyer, 1887.

Rowland's mapping of the solar spectrum was based on photography. During the 1880s, the early spectroscope, using visual observation of spectra, was becoming the spectrograph, with photographic recording. The value of such recording derives in part from the ability of the photographic plate to accumulate the effects of incident light, so that long exposures can reveal details too faint to be seen with the naked eye, but nineteenth-century astrophysicists recognized a number of other advantages. The scintillation of light sources made visual spectroscopy impossible in bad seeing, whereas photography was little affected. The human eye required a greater width of visual, than of photographic, spectrum in order to distinguish lines. Finally, wavelengths could be measured much more accurately from photographs in a laboratory than at the eyepiece of a telescope. Initially, these advantages were offset by the low sensitivity of photographic plates. This did not retard observations of the solar spectrum (which was first photographed in the 1840s), but photography of stellar spectra became commonplace only in the last two decades of the nineteenth

century. Correspondingly, the most sophisticated piece of spectroscopic instrumentation developed in the nineteenth century – the spectroheliograph – was concerned with photography of the Sun.

The detection of solar prominences out of eclipse, by observing in an isolated characteristic emission line, began in the latter part of the 1860s. Such work was pursued throughout the 1870s and 1880s by various observers, especially in Italy, but the standard method of recording continued to be drawing from visual observation. At the end of the 1880s, George Ellery Hale (1868–1938) devised his spectroheliograph to record prominences (and the chromosphere) photographically. The principle of the instrument had been suggested more than once before, but it had not previously been implemented successfully; though, almost simultaneously with Hale, H. A. Deslandres was developing an instrument with the same intention, but of different construction. Hale's original version employed two stationary slits – the first to feed into the spectroscope and the second to isolate the required emission line. The solar image was moved across the first slit and the photographic plate

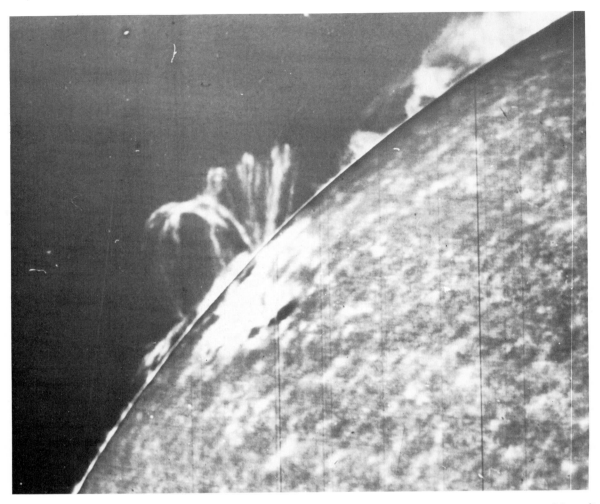

4.3. Philip Fox's montage of a spectrogram of a prominence and a spectroheliogram of the underlying spots and flocculi on the disc, 14 August 1907. Both Yerkes photographs were made in the light of ionized calcium.

simultaneously across the second, so that a monochromatic image of the prominence could be constructed. This instrument was only partially successful, and was replaced by an alternative scheme that kept the solar image and photographic plate stationary while moving the slits. More fundamentally, Hale reconsidered the process of photographically recording emission lines. He had originally chosen a hydrogen line for his observations – a traditional choice for examining prominences. However, photographic plates were more sensitive in the ultraviolet, so Hale next turned to the H and K lines of calcium. These were sufficiently far in the ultraviolet to be difficult to observe visually, but Hale noted that their bright

centres in the middle of very wide dark absorption lines made them ideal for photographing prominences with a minimum of background interference.

When the new spectroheliograph came into operation in 1892 it proved an immediate success, but it did more than transform prominence observations. Studies of both prominences and the chromosphere, whether at eclipses or otherwise, had always been carried out at the solar limb, and this was also the intention with the spectroheliograph. But Hale noted that, when his instrument scanned the disc of the Sun, distinct clouds – which he labelled 'flocculi' – became visible. The spectroheliograph thus provided a means of examining

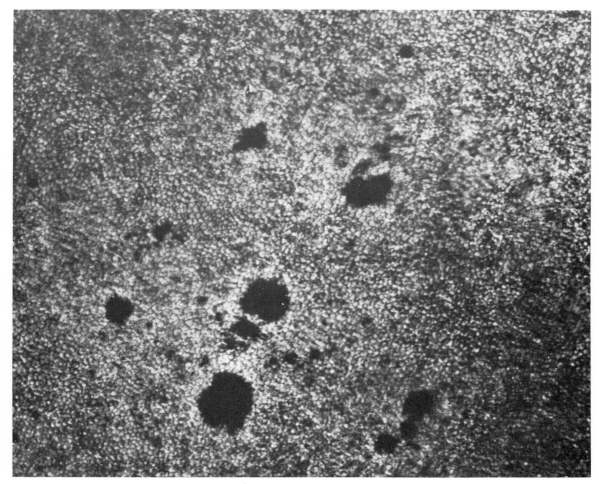

4.4. Solar granulation, pores, and small spots, 5 July 1885. Throughout the first half of the twentieth century this photograph of solar granulation taken at Meudon by Jules Janssen remained the best one in existence. Janssen wrote that it was "obtained without intervention by the human hand".

the solar atmosphere both at the limb and across the disc. It rapidly gained popularity: Hale recorded in 1908 that spectroheliographs were in daily use in India, Sicily, Spain, Germany and England, as well as in the United States.

Hale's use of the properties of photographic emulsions to enhance the capabilities of his instrumentation was not unique. In the 1870s, P. J. C. Janssen made an extensive study of the way in which photographic emulsions reacted to very short exposures to sunlight. As a result, he managed to produce photographs of the solar granulation that were not surpassed until after World War II. During the 1870s, an effort was made, especially in France, to determine the shortest wavelength of

solar radiation that could be recorded photographically. (A. Cornu ultimately claimed a value of $\lambda = 2930$ Å.) At the same time, attempts were under way to extend photography at the red end of the spectrum. Although the ordinary photographic plate was naturally most sensitive in the violet, it was recognized that the range of sensitivity could be changed by the addition of appropriate dyes to the emulsion. In this way, by 1880, W. de W. Abney in England had extended photography of the solar spectrum to more than one micron. (These ultraviolet and infrared limits may be compared with the limits of the visual spectrum, which, from the 1870s onwards, were generally quoted as from about 3900 to 7600 Å.) But the biggest extension of the spectrum in the nineteenth century was not due to improvements in photography: it derived from Langley's development of the bolometer during the last two decades. With this instrument, he observed the solar infrared spectrum out to 5.3 μm. His bolometer consisted essentially of two strips of platinum, situated side by side and forming the two arms of a Wheatstone bridge: one of the strips was exposed to the solar spectrum, while the other was shielded from it. The variations in reading along the spectrum (produced via a rock-salt prism for the longest wavelengths) were recorded by photographing the consequent galvanometer deflections. Absorption bands could only be traced in detail over part of this total spectral region – which Langley enthusiastically dubbed 'the new spectrum' – but it was clear that infrared absorption was greatly affected by the terrestrial atmosphere.

The effects of the Earth's atmosphere on astronomical observation became a matter for increasing concern during the latter part of the nineteenth century. The idea of making observations at high altitudes to cut down these effects was not new, even then, but the need for highly detailed solar observations triggered off a series of attempts to establish observatories, either permanent or temporary, on mountain tops. Much of this effort was undertaken in the USA where there were numerous suitable sites from which to choose. But interest in the possibilities was spread more widely; as is indicated, for example, by Janssen's establishment of a solar observatory on Mont Blanc in the 1890s. The first important astrophysical expedition to high altitudes was probably that mounted in the United States by C. A. Young (in 1872) to carry out detailed studies of the chromosphere. Young detected many more emission lines than he could find at sea-level. Nearly ten years later, Langley went to Mt Whitney in California to carry out his observations of the solar infrared, and to measure the solar constant. His visit had the additional aim of determining whether it would be feasible to set up a permanent high-altitude station. The founding of a solar observatory on Mt Wilson in 1904 can be seen as the first consequence in the twentieth century of this nineteenth-century concern. Mt Wilson had, in fact, been favoured earlier by E. C. Pickering, but Harvard's main interest in high-altitude observing concentrated on Peru. The factors involved in these nineteenth-century discussions of the siting of observatories have had a major impact on the development of new observatories in the twentieth century.

Problems of nineteenth-century astrophysics

A basic element that prevented more rapid progress in nineteenth-century astrophysics was the difficulty in interpreting observations. This can be illustrated by a number of examples, beginning with the measurement of surface temperatures.

We noted that Langley's high-altitude observations were partly aimed at determining the solar constant. This latter was of considerable interest at the time, both for the extent to which it was really constant, and for the information it could give on the surface temperature of the Sun. The former question proved difficult to answer owing to major differences between the various determinations. French measurements in the 1870s increased the previously accepted value of $1\frac{3}{4}$ cal cm^{-2} min^{-1} to $2\frac{1}{2}$. Langley subsequently increased this to 3. But estimating the solar surface temperature was held up by a theoretical problem: how to relate the amount of radiation emitted by a surface to its temperature. For much of the nineteenth century two methods, based on laboratory measurements, competed for acceptance – Newton's law and Dulong and Petit's law. Both were used for deriving the surface temperature of the Sun from the solar constant: in terms of our present-day knowledge, Newton's law always gave results that were too high, and Dulong and Petit's law results that were too low. An indication of these differences is given in *The New Astronomy*:

It is probable, from all experiments made up to this date, that the solar effective temperature is not less than 3000 nor more than 30000 degrees of the centigrade thermometer.

Some five years before Langley wrote this, Josef Stefan had already proposed the relationship between radiation and temperature accepted today – that the dependence goes as the fourth power of the temperature. It was some time, however, before this relationship was accepted, for it was not clear why it should be preferred to other suggestions. (For example, a dependence on the square of the temperature was suggested in the same year that Stefan published his own proposal.) Other methods put forward, such as estimating the surface temperature from the line spectrum or from the peak intensity of the continuous spectrum, also produced a variety of possible surface temperatures. This uncertainty about the Sun affected determination of all stellar surface temperatures, since ultimately the latter required calibration in the same terms. By the end of the century, cautious agreement was being reached on a temperature not too far from that accepted today, but this agreement needed the underpinning of twentieth-century radiation theory before it could become firmly established.

The question of the surface temperature of the Sun necessarily influenced discussions on the nature of the solar photosphere. The well-defined surface of the Sun was taken as an indication that the photosphere consisted of clouds of some kind of particulate matter, and this seemed to be confirmed by spectroscopy. As J. Scheiner wrote in *Die Spectralanalyse der Gestirne* (1890):

We might assume that the density of the metallic vapours in the deeper layers is so great that their lines are broadened until the spectrum becomes continuous, and that the upper, less dense and cooler portions of these same masses of vapour produce the selective absorption. But for this to be true, the appearance of the absorption lines would have to be different from what it really is, for it would be impossible that the majority of the lines should be so sharp, since the transition from the emitting to the absorbing layers would necessarily be continuous. We are therefore forced to conclude that the light of the photosphere is due to solid or liquid particles.

The problem was to decide on the nature of the cloud particles; for even the lowest estimated solar temperatures were sufficient to vaporize most substances known on Earth. One possibility was carbon, which was both refractory and, by the 1880s, known to be present in the solar atmosphere. But, in that case, what would be the nature of the photosphere in stars much hotter than the Sun? These questions were raised, but did not become pressing until a reasonable scale of surface temperatures had been fixed. Then, in the twentieth century, the explanation came from developments in the theory of radiative transfer.

Besides the need to explain the nature of the photosphere, nineteenth-century astrophysicists were faced with the problem of understanding the varied phenomena of solar activity. The theoretical basis here – magnetohydrodynamics – was not properly developed until after World War II, but some hint of the right direction is already discernible in nineteenth-century studies of sunspots. The main difficulty with spots lay in understanding what caused the absorption, making them darker than the rest of the solar surface. At first, it was thought the cause might be an accumulation of the particles that were bright elsewhere in the photosphere. (The 'Wilson effect', which was much discussed during the last decades of the century, pictured sunspots as depressions in the solar surface.) But, in 1883, Young showed that spot spectra seen under high dispersion actually consisted of many fine absorption lines close together. N. C. Dunér later confirmed this, adding that the continuous spectrum between the lines differed little in brightness from the photospheric continuous spectrum. The unavoidable conclusion from these observations was that the extra absorption in sunspots was due to gases, not to solids or liquids. Young also noted that the spectral lines in spots were often 'spindle-shaped', rather than the same width across the entire spot. This observation ultimately proved to be the clue to the magnetic properties of sunspots, but, in the nineteenth century, was attributed either to increased pressure in spots or to the occurrence of chemical reactions there.

There remained the question – why do spots form in the first place? The most widely accepted answer reflects the typical nineteenth-century picture of the solar atmosphere as being understandable in the same terms as terrestrial meteorology. Spots were seen by nineteenth-century astrophysicists as cyclonic motions in the cloudy photosphere of the Sun, analogous to similar motions in the Earth's atmosphere. The major difficulty was that only a few spots showed definite signs of vorticity when observed visually, although Hale believed later that his spectroheliograph observations of the chromosphere near spots provided clearer evidence. He seemed justified in the early twentieth century when he used the concept of cyclonic motions to provide an explanation of sunspot magnetism.

Sunspots were supposed to hold two main

lessons for stellar astrophysics. The first related to stellar variability: when an explanation of variability in terms of eclipsing binary behaviour appeared to be improbable, the normal alternative offered was obscuration of the surface by spots. The second concerned sunspot spectra, which were found to be similar to those of some red stars. As Scheiner explained, these two points were linked:

Bands similar to those in spectra of Type IIIa have also been found to be present in spot spectra, so that quite a marked resemblance can be traced between the spectrum of sunspots and that of stars of this class. All this goes to show that these stars represent a further stage in the development in the direction of lowering temperature and increasing density, and we should accordingly imagine the greater part of their surface to be in a condition similar to that of sunspots – whence we obtain an explanation of the fact that so many stars of this class are either irregular or long-period variables.

Many of the problems of interpretation in nineteenth-century astrophysics concerned spectra. There were two distinct types of difficulty. The first stemmed from the uncertainty about surface temperatures, and equal uncertainty as to the way in which both temperature and pressure affected spectral lines. The second followed from the lack of agreement concerning the basic nature of matter and, more particularly, from the inability to explain the processes by which spectral lines were produced. The former problem is well illustrated by nineteenth-century attempts to interpret spectra at different levels in the solar atmosphere. By analogy with the Earth's atmosphere, it was supposed that both temperature and density in the Sun's atmosphere decreased with height. Lockyer spoke for most astrophysicists when in 1887 he painted the following picture in *The Chemistry of the Sun*:

1. We have terrestrial elements in the sun's atmosphere.
2. They thin out in the order of vapour density, all being represented in the lower strata . . .
3. In the lower strata we have especially those of higher atomic weight, all together forming a so-called 'reversing layer' by which chiefly the Fraunhofer spectrum is produced.

Despite the general agreement, this model raised

a number of problems. For example, the H and K lines of calcium were known to be strong in the chromosphere; but calcium was a relatively heavy atom, and should therefore have been concentrated lower down in the atmosphere. Perhaps the most striking difficulty, however, concerned the coronal spectrum. The first coronal line observed (usually labelled '1474K', because this was its estimated wavelength on the scale Kirchhoff originally introduced) seemed readily identifiable. It appeared to be coincident with a known iron line found in both photospheric and chromospheric spectra. This obviously raised the query how such a heavy atom could rise to such great heights; but the problem was soon complicated by the discovery of other coronal lines, not attributable to iron or, indeed, to any other known element. Hence, the suggestion arose that part of the coronal spectrum might be produced by some element yet to be discovered on Earth. This conjecture seemed to be strengthened at the end of the century, when it was found (at the 1898 eclipse) that the wavelength of the 1474K line had been incorrectly measured at earlier eclipses: it was not coincident with the iron line. In view of the assumptions we have listed about the solar atmosphere, the obvious conclusion was that such a new element – 'coronium' – would have to be lighter than hydrogen in order to spread out so far from the Sun. It was only gradually realized during the twentieth century that this entire picture was wrong; basically because, when the spectra of the chromosphere and corona could be interpreted, they revealed that the temperature in the solar atmosphere increased, rather than decreased, with height.

The difficulty of deciphering spectra in the nineteenth century derived not just from the complexity of the spectrum of each element, but from apparent changes in spectra when they were produced in different ways. Various attempts were made to explain these changes (or to explain them away); the most interesting, and most discussed, being based on the proposition that the ultimate atoms making up a gas might be broken down further, under sufficiently extreme conditions, to produce a new kind of spectrum. Unfortunately, it proved very difficult to turn this vague statement into an acceptable model. (There was even a confusion in the terminology, since physicists and chemists tended to use the word 'molecule' in different senses.) These difficulties are very apparent in the most detailed scheme of atomic breakdown devised in the nineteenth century – Lockyer's dissociation hypothesis. This was, indeed, originally based on a complete misconception. It was noted in the early days of spectroscopy that different elements often appeared to have some of their lines coincident (i.e. situated at the same wavelength). Lockyer called these 'basic' lines, and believed they reflected the presence of a common breakdown product of the initial elements. However, as higher dispersion instruments became available, it was found that the supposed common lines were merely near to each other in wavelength, not coincident, and therefore possessed no fundamental significance.

Towards the end of the century, the evidence from the spectra of white stars led Lockyer to alter the form of his dissociation hypothesis. He now called attention to a different group of lines, the 'enhanced' lines, claiming that these indicated the existence of atoms that had been dissociated. Here he was essentially looking forward to what was subsequently labelled 'ionization', but his reformulation led to too much confusion for general acceptance at the time. The concept of subatomic particles was by this time beginning to receive support from laboratory experiments, especially by J. J. Thomson, and it was via such work that an adequate understanding of ionization was finally reached in the twentieth century. During the nineteenth century, many observations of astrophysical spectra were necessarily explained either inadequately, or incorrectly. Scheiner explained in *Die Spectralanalyse der Gestirne* why he was not prepared to discuss observations of the solar spectrum in detail:

The spectroscopic investigations upon the Sun are so closely associated with theories of its constitution that it is impossible to give a detailed account of the investigations without a thorough discussion of the theories. But such a treatment would require a whole book for the Sun alone, which is an adequate reason for its omission here. . . . There was another and deeper reason, however, for deciding against a complete discussion. It is impossible to overlook the fact that the present state of our knowledge as to the constitution of the central body of our system does not satisfy the expectations that might be fairly entertained. On the one hand, we have available a large mass of observational data, although for the most part unscientifically discussed, and on the other hand an indefinite number of hypotheses and solar theories, which are with few exceptions radically wrong at the start and often contradictory to the most simple physical views of today.

There was one area of spectroscopy where lack of a theoretical basis did not hinder progress – the measurement of radial motion via the Doppler effect. Because Doppler shifts were small, and therefore required high dispersion spectra, they were first detected on the Sun. Initially, there was some doubt whether the observations could be interpreted in terms of radial motion, since some of the measured velocities in the chromosphere were much higher than expected. The applicability of the Doppler principle to the solar spectrum was ultimately confirmed in 1871 by H.C. Vogel – by spectroscopic measurement of the solar rotation rate, which had already been determined from visual observation of spots. Spectroscopic measurements of solar rotation, and especially of the differential rotation, continued throughout the nineteenth century. Dunér once commented that he thought the explanation of the results was one of the most difficult problems in astrophysics.

Reliable studies of stellar radial velocities awaited the introduction of photography: low dispersion spectra could only be measured with sufficient accuracy under a microscope. The first major investigation in this area was begun by Vogel at Potsdam towards the end of the 1880s, and was soon followed by a long-continued campaign at the Lick Observatory in California. These measurements played their part in reconciling classical astronomers to the new astronomy, for the radial motions derived spectroscopically could be combined with proper motions obtained by classical methods to yield the space motions of stars. Equally, the well-established interest of classical astronomers in the motions of visual binaries could be matched by the new observations of spectroscopic binaries. These latter became a major research topic in the late nineteenth century, beginning at the end of the 1880s with Pickering and Vogel, and continuing in the 1890s with A. A. Belopolsky at Pulkovo and W. W. Campbell at Lick. The number of catalogues containing astrophysical data by the end of the century bears witness to the increasing assimilation of astrophysical observations into the classical tradition.

The growing importance of astrophysics

The increasing role of astrophysics in astronomical research during the latter part of the nineteenth century can be discerned in the journals of the period. Traditional astronomical publications, such as the *Astronomische Nachrichten* in Germany or the *Monthly Notices of the Royal Astronomical Society* in Britain, carried an increasing proportion of astrophysical contributions as the century progressed. In addition, the various observatory publications, which were then very common, contained much important astrophysical research: the Harvard College Observatory *Annals* under E.C. Pickering are an obvious example. But the most interesting development was the creation of new journals specifically devoted to astrophysics. A pioneer in this respect was the *Memorie della società degli spettroscopisti Italiani*, which first appeared in the early 1870s and covered both astronomical and laboratory spectroscopy. This mixture of the two topics has become less common in the twentieth century, but equally characterized what became the most important of the new publications – *The Astrophysical Journal*.

In October 1891, Hale wrote to Pickering:

After consulting with Prof. Young, Dr Huggins, Dr Vogel, and others, I have decided to publish "The Astro-Physical Journal" in place of ordinary publications from this observatory [the Kenwood Physical Observatory] . . . I propose to publish the journal at irregular intervals, but as often as sufficient material is at hand, and it seems desirable to translate a good many articles from foreign journals, in order to collect under a single cover a large proportion of the literature relating to spectroscopy and allied subjects.

Hale actually encountered resistance to the idea of a journal totally devoted to astrophysics. Instead, a compromise was worked out whereby Hale acted as editor of a new astrophysical section in an existing journal, the *Sidereal Messenger*, the composite result being retitled *Astronomy and Astro-Physics*. The union did not last long: the astrophysical section attained sufficient viability to split off in 1895 to form *The Astrophysical Journal*, an

International Review of Spectroscopy and Astronomical Physics. The initial difficulty in establishing the journal indicates that the relative status of astronomy and astrophysics remained a matter for debate in the 1890s. This is also suggested by the discussions leading to the creation of the American Astronomical Society. Two successful conferences were held in the USA in 1897 and 1898 to bring together astronomers and physicists. As a consequence, the carefully named Astronomical and Astrophysical Society of America was established in 1899. Despite the clumsiness of this title, it was felt expedient to retain it for the next fifteen years, before general agreement was reached that both branches of the subject could be covered by the one word 'Astronomical'.

The growth of astrophysics during the last three decades of the nineteenth century may seem inevitable in retrospect; at the time, participants saw it more as a struggle for the necessary support. Classical astronomy could, at least, claim to have its uses. Astrophysics seemed to have no such claim. As Huggins said:

The new astronomy, unlike the old astronomy to which we are indebted for skill in the navigation of the seas, the calculation of the tides, and the daily regulation of time, can lay no claim to afford us material help in the routine of daily life.[4]

Why then was astrophysical research expanded so successfully in the nineteenth century? Much of the work was carried out at private or university observatories, where usefulness was not a requirement. Nevertheless, it seemed for a time in the nineteenth century that some astrophysics might ultimately be justified on the grounds of practical application. The study of solar–terrestrial relations, which began in the mid-nineteenth century by correlating terrestrial magnetic variations with the sunspot cycle, led later into a detailed examination of possible links between the solar cycle and terrestrial meteorology. For some years much was hoped and claimed for this work, and it was correspondingly argued that solar physics deserved not merely recognition, but also financial aid. This aspect of weather prediction was still being strongly pursued at the beginning of the twentieth century, when Langley wrote to Lockyer, one of its chief proponents:

It is an amazing thing to me that the enormous

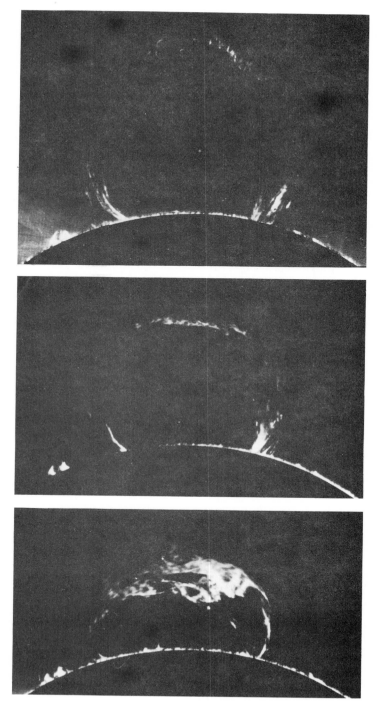

4.5. Edison Pettit's calcium H-line pictures of the great prominence of 15 July 1919. The three views show the increase in height of 200000 km in 1 hour. Made with the 40-inch refractor and Rumford spectrograph at Yerkes Observatory.

utility *of such work as yours on the Sun's connection with the conditions which bring famine or plenty, to India, for instance – the immense* utility *of like studies is lost sight of by almost all astronomers.*[5]

As Langley's letter hints, most astronomers were much less enthusiastic supporters of a link between solar activity and terrestrial weather. Although the importance of the Sun for life on Earth continued to be stressed in a general way, most requests for support appealed rather to national, or personal, pride and to the intrinsic fascination of the subject. When Joseph Winlock asked the US Congress for funds to finance an eclipse expedition in 1869, he justified his appeal by saying:

In 1843 the United States held a low rank among the Nations in respect to the Exact Sciences . . . In 1869 all is changed . . . We must not allow her [the USA] to forfeit this high position.

Hale, the greatest of all fund-raisers for astrophysics, commenting on his own efforts in an influential article in *Harper's Magazine* in 1928, remarked:

I have had more than one chance to appreciate the enthusiasm of the layman for celestial exploration. Learning in August, 1892, that two discs of optical glass, large enough for a forty-inch telescope, were obtainable through Alvan Clark, I informed President Harper of the University of Chicago, and we jointly presented the opportunity to Mr Charles T. Yerkes. He said he had dreamed since boyhood of the possibility of surpassing all existing telescopes, and at once authorised us to telegraph Clark to come and sign a contract for the lens.

These two examples have been quoted from the experience in the USA because, in the latter part of the nineteenth century, new developments in astrophysical telescopes, instrumentation and observation occurred most strikingly in that country. It is noticeable how European astronomers increasingly appealed to American astronomers for assistance. For example, Huggins during a prolonged dispute with Lockyer over the source of the chief emission line in nebulae repeatedly appealed to J.E. Keeler at Lick for observations, and it was

Keeler's results that were finally conclusive. The growing American dominance of observational astrophysics was not simply due to the possibilities of funding or to the good observing conditions, but, behind these, to the acceptability of the subject in the United States. Astrophysics in the nineteenth century required practical ingenuity and a willingness to travel, without need for a deep involvement in mathematics. Its results could often be presented pictorially (in contrast to the more mathematically based classical astronomy), a form guaranteed to fascinate the general public. Moreover, astrophysics was a new area of investigation: whereas Americans came to classical astronomy when it was already well established, they could argue that they were helping to create astrophysics. For all these reasons, the growth of astrophysics in the nineteenth century was paralleled by the growing influence of the United States in astronomy. As Huggins wrote to E.S. Holden in the 1870s: "You Americans have 'l'avenir du Monde'."

Notes

1. W. Huggins, *Publications of Sir William Huggins's Observatory*, vol. 1 (London, 1899), p. 22.
2. O.A. Melnikov, Towards a history of the development of astrospectroscopy in Russia and the USSR (in Russian), *Istoriko-Astronomicheskie Issledovania*, vol. 3 (1957), 9–258, quotation from p. 27.
3. Letter to an unidentified correspondent, 10 October 1890; quoted in B.Z. Jones, *Lighthouse of the Skies* (Washington, DC, 1965), p. 124.
4. W. Huggins, The New Astronomy: a personal retrospect, *The Nineteenth Century*, vol. 41 (1897), 907–29, quotation from p. 907.
5. Report by the Director of the Solar Physics Observatory [Cd 5912] (1911), p. 17.

Further reading

Agnes M. Clerke, *Problems in Astrophysics* (London, 1903)
Bessie Zaban Jones, *Lighthouse of the Skies* (Washington, DC, 1965)
Bessie Zaban Jones and Lyle Gifford Boyd, *The Harvard College Observatory* (Cambridge, Mass., 1971)
William McGucken, *Nineteenth-Century Spectroscopy* (Baltimore and London, 1969)
A.J. Meadows, *Early Solar Physics* (Oxford, 1970)
A.J. Meadows, *Science and Controversy: A Biography of Sir Norman Lockyer* (London, 1972)
Helen Wright, *Explorer of the Universe: A Biography of George Ellery Hale* (New York, 1966)

5

Variable stars

HELEN SAWYER HOGG

The early visual discoveries of variable stars were made sometimes by chance and sometimes by long and tedious comparisons. A new dimension was added to the search when photographic plates of star fields became available in the later decades of the nineteenth century, and before long the numbers of known variables shot upwards. The enthusiasm of F.W.A. Argelander (1799–1875) in Europe for the discovery and investigation of variables was matched years later by that of E.C. Pickering (1846–1919) on the other side of the Atlantic. And the Harvard College Observatory began to acquire resources never before available for astronomical studies, a rapidly increasing treasure of photographic plates, in time amounting to tens of thousands, taken with its many telescopes in both the northern and southern hemispheres.

As the number of variable stars increased (Figure 5.1), the first nomenclature adopted by Argelander in 1844 proved inadequate. This used for each constellation the letters R, . . ., Z and so could be applied to only nine variables per constellation. In 1881 the double letter system was introduced by C.E.A. Hartwig (1851–1923), RR, RS, . . ., RZ, SS, ST, . . ., SZ, TT, . . ., ZZ. A later extension through AA, AB, . . ., AZ, BB, . . ., QZ permitted 334 combinations in all (J not being used). For example, from a Harvard photograph taken on 13 July 1889 with 13 exposures of 29^m 40^s each, Williamina P. Fleming (1857–1911) found that star BD +42° 3338 varied just under a magnitude; this variable became RR Lyrae. Before the letters were assigned, however, confirmation of the variability had to be obtained from the Committee on Variable Stars of the Astronomische Gesellschaft and a provisional number assigned by the Editor of the *Astronomische Nachrichten*. However, at the Harvard College Observatory the

variables found there in large quantities were numbered consecutively, and these designations are still sometimes used.

In the twentieth century the enlarged alphabetical system proved inadequate and a suggestion made in 1899 by C. André, director of the Lyons Observatory, was eventually adopted by the International Astronomical Union in 1925. After the alphabetical terminology is exhausted, variables are designated by the letter V followed by arabic numerals in order of discovery in the constellation. For example, QZ Carinae is followed by V335 Carinae.

From the mid-nineteenth century on, encouraged to a great extent by Argelander's enthusiasm, more and more observers participated in the search for variables and observations became increasingly numerous, while the photometers in use gained in accuracy. Argelander used a step method of estimating the brightness of a variable among comparison stars; later, as magnitudes

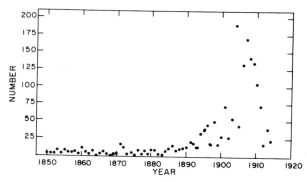

5.1. The number of variables discovered per year for the 1689 stars catalogued by G. Müller and E. Hartwig up to 1915. The upswing at the turn of the century reflects the impact of photography; the sudden decline shows the influence of the war years in slowing both the discovery and the verification of variable stars.

were determined for an increasing number of comparison stars, estimates were made directly in tenths of magnitudes. By the last decade of the nineteenth century visual observations had piled up by the thousands; but many of these were unpublished, leading to concern that they might be lost forever. Beginning in 1884 Pickering published in the *Proceedings* of the American Academy of Arts and Sciences lists of current observations of variable stars. In 1890 in the Harvard *Annals* he extended the index to cover the entire period from 1840 to 1887, mainly with unpublished observations, listing the number of observations per star available from the work of more than thirty observers. Later he worked to get into print the actual observations deemed especially useful; for example, in 1900 Pickering published in the Harvard *Annals* those made between 1845 and 1879 by J.F.J. Schmidt, director of the Observatory at Athens.

Argelander had realized that only a cooperative effort of many observers could achieve as complete a record as possible of the light changes of a variable star, and he considered the observation of variable stars an ideal pursuit for amateur astronomers. His pamphlet "An appeal to the friends of astronomy" in 1844 in H.C. Schumacher's *Jahrbuch* was the basis of Pickering's bulletin in 1882, "A plan for securing observations of variable stars". Pickering thought the idea would appeal particularly to women.

Certain types of variables were especially suited for study by amateurs with small telescopes. Long-period variables have large ranges and do not repeat their light curves exactly from cycle to cycle, and hundreds of them are bright enough to be seen with small telescopes. Also suitable are the few R Coronae Borealis and U Geminorum stars, whose behaviour is both exciting and unpredictable. In 1890 a Variable Star Section was established by the British Astronomical Association (BAA), the members' observations being published in the Association's *Journal* and *Supplements*. In 1911 the American Association of Variable Star Observers (AAVSO) was started at the Harvard College Observatory. Argelander's appeal of 1844 was brought to the fore when Annie J. Cannon (1863–1941) recalled it to the new AAVSO, and contrasted the knowledge of variables at these two widely separated years. Instead of the 18 variables

known in 1844, in 1911 there were 4000; instead of observers in "Aachen, Breslau and in Bonn", by 1911 they were to be found in nearly every country in Europe, nearly all of the United States, in Japan, in South America, in Australia, in Egypt and in South Africa. Much of our understanding of certain types of variables comes from the persistence and remarkably long continuous records of the BAA and AAVSO observers, and societies in other countries have followed their example.

Star charts are an essential tool in the visual study of variables. The early charts were developed mainly from Argelander's *Bonner Durchmusterung*. They culminated in the work of the Father J.G. Hagen, of the Georgetown College Observatory, completed at the Vatican Observatory 1899–1908: the *Atlas Stellarum Variabilium*, in six series. As the amateur societies were formed, they often developed their own extensive charts.

Though the first photograph of a star was obtained in 1850, it was not until the last fifteen years of the century that celestial photography had much effect in providing observations of variables. Several techniques were used in systematic searches. If a series of exposures was taken on the same plate on a single night, a variation in a star's image could be very conspicuous; and the same might be true if positive and negative photographs of a region taken on different dates were superposed, or two positive photographs examined in a stereocomparator or blink microscope. Not only did photography substantially increase the ease of discovery of new variables, but it had a further impact in two ways on their observation. Dozens, and occasionally hundreds, of variables could be studied on each photographic plate. Furthermore, such a record is of a semi-permanent nature. No longer was the first observer of an alleged variable limited to reporting a fleeting impression of the star, which no one else might corroborate. The existence of a plate made it possible for an independent observer to check the brightness of a variable on the plate years or even decades after the first estimate of its magnitude.

In 1786 Edward Pigott had brought together what was then known of variable stars in his *Philosophical Transactions* catalogue of 12 confirmed and 38 suspected variables entitled "Observations and remarks on those stars which the astronomers of the last century suspected to be

5.2. The organizational meeting of the American Association of Variable Star Observers at Harvard College Observatory, 10 November 1917. The women seated in front row are Leah Allen, Annie J. Cannon, and Henrietta S. Leavitt. In the centre of the back row are E.C. Pickering and Solon I. Bailey.

changeable". In the nineteenth century, beginning with Argelander in 1844, there appeared one small catalogue of variable stars after another, the largest being S.C. Chandler's list of 393 variables published in *The Astronomical Journal* in 1896. From 1870 the "Katalog und Ephemeriden der Veränderlichen Sterne" was published annually in the *Vierteljahrsschrift* of the Astronomische Gesellschaft until 1926, when R. Prager began to publish it as *Kleinere Veröffentlichungen der Universitäts Sternwarte zu Berlin-Babelsberg*. Lists of variable stars were also published in Paris in the *Annuaire du Bureau des Longitudes*.

The intense data-gathering drives of the late nineteenth century, which resulted in large star catalogues and in catalogues of thousands of nebulous objects and clusters of stars, also had

their effect on the rate of discovery of variable stars. By the early twentieth century over one thousand variable stars were known and listed in the first *Harvard Provisional Catalogue* of 1903. This was followed in 1907 by the *Second Catalogue of Variable Stars* by Miss Cannon under the direction of Pickering, with a later supplement containing all the variables discovered to 10 December 1909. These contained a total of 1425 variables and 23 new stars (with further reference to 551 variables in 22 star clusters). Extensive bibliographies were a prerequisite to cataloguing the variable stars. At Harvard the bibliography was begun in 1897 by W.M. Reed, who wrote 15000 cards, before the task was transferred to Miss Cannon, who added 20000 prior to publication of the *Second Catalogue*. The culmination of the variable star catalogues of this period was that by G. Müller of Potsdam and Hartwig of Bamberg, *Geschichte und Literatur des Lichtwechsels der Veränderlichen Sterne*, published at Leipzig in three volumes between 1918 and 1922 giving the elements and all references in the literature for over 1700 stars.

In addition to variable stars in the Galaxy, variables were being identified in clusters and in the Magellanic Clouds, and these were to prove of the greatest significance. The first evidence of variability in a globular cluster was the nova discovered in M 80 in Scorpius in 1860. It reached at least magnitude 7.0, bright enough to change the appearance of the cluster. It was found by A. Auwers on 21 May, confirmed by E. Luther, and seen independently by N. Pogson on 28 May. The last sighting was on 16 June by Auwers at magnitude 10.5. (This object is still the only nova to have been seen visually in a globular cluster.)

Three decades elapsed before other variables in globular clusters attracted attention. Then, in 1889–90, independent observations by Pickering, A. A. Common, E. E. Barnard and D. E. Packer led to the discovery of the three apparently brightest variables in northern hemisphere globular clusters, one star in M 3 and two in M 5. In the next decade more than 500 variables were found in globular clusters from Harvard photographs taken at Arequipa, Peru. The bulk of such research was carried out by S. I. Bailey and his associates, resulting in his classic Harvard *Annals* volume in 1902, with its catalogue of 128 variables in Omega Centauri, and identifications for 395 variables in 16 other globular clusters. By 1930 nearly 900 variables were known in 42 clusters.

By the end of the first decade of the twentieth century the number of variables in nebulae had become comparable with the number known as isolated stars in our Galaxy. It was the Magellanic Clouds that yielded variables in large numbers, an unexpected discovery due to astronomers at Harvard College Observatory where variable star enthusiasm was at a peak as a result of the finding of hundreds of variables both in globular clusters and in the galactic fields. The discoveries in the Clouds began in the spring of 1904 when comparison of two photographs of the Small Magellanic Cloud taken with the 24-inch (61-cm) Bruce refractor at Arequipa, Peru, led to the detection of a number of faint variable stars. A series of 16 plates

taken that autumn for period determination showed "an extraordinary number of new variable stars" when examined in January 1905 in Cambridge, Massachusetts. In 1908 Henrietta S. Leavitt published her classical paper, "1777 variables in the Magellanic Clouds", and most of them later proved to be Cepheids, with important consequences to which we shortly come.

Pickering's classification

To understand the nature of the star's variability, determination of the period of its light changes is usually necessary. The difficulty of determining a period varies with the magnitude, range, period length and number of observations available, and is in itself a subject for treatises.

As the number of variables with known periods increased, a pattern of types became obvious. Three types were recognized early and Pigott used these as a classification in 1786: long-period variables, new stars and short-period variables. The first two were fairly clear cut in their characteristics, but the short-period variables were a mixed bag of stars with different causes for their variations, and some of these stars were to prove most puzzling. In many cases the solution to the puzzle was to come from spectroscopy.

The classification having the widest acceptance in our period was that proposed by Pickering in 1880 to the American Academy of Arts and Sciences, and revised by him in 1911. In what follows, variables will be discussed in the order of Pickering's 1911 system, namely:

(1a) Normal novae: Nova Aurigae 1891, Nova Persei 1901;

(Ib) Novae in nebulae: Nova Andromedae 1885, Nova Centauri 1895;

(IIa) Usual long-period variables: Omicron Ceti, Chi Cygni;

(IIb) U Geminorum type: U Geminorum, SS Cygni, SS Aurigae;

(IIc) R Coronae Borealis type: R Coronae Borealis, RY Sagittae, SU Tauri;

(III) Irregular variables: Alpha Orionis, R Scuti;

(IV) Usual short-period stars: Delta Cephei, Zeta Geminorum (these stars originally formed class IVa; class IVb, Beta Lyrae type, was moved in 1911 to class V);

(V) Variables of the Algol type: Beta Persei, Delta Librae.

Pickering took no notice in this classification of the important distinction Bailey had made in 1902 when he found large numbers of variables that he termed 'cluster type' in the globular cluster Omega Centauri. Bailey proposed that Pickering's class IV (which included all short period variables except Algol stars) should be subdivided into three subclasses. Subsequently the cluster type variables came to be recognized as having the same characteristics as those variables called RR Lyrae type in galactic fields.

Normal novae, Type Ia

The light curves of novae, more than those of any other variables in galactic fields, depended on photography. Table 5.1 lists the novae that were identified before the twentieth century.

The light curves of novae showed a rapid rise

Table 5.1. *Novae recognized in the West before the twentieth century*

Year of appearance	Name	Apparent magnitude	Discoverer
1572[+]	B Cas	−5	Schuler
1600	P Cyg	3.5	Blaeu
1604[+]	N Oph 1	−4	Brunowsky
1670	11 Vul	3	Anthelme
1783	N Sge	6	D'Agelet
1827[*]	η Car	1	Burchell
1848	N Oph 2	5.5	Hind
1854	N Ari	9.5	Krueger, Chacornac
1860	T Sco 1	7.0	Auwers, Pogson
1860	T Boo	9.7	Baxendell
1862	N Ara	5	Tebbutt
1863	N Sco 2	9.1	Pogson
1866	T CrB	2.0	Birmingham
1876	Q Cyg 2	3	Schmidt
1877	N Com	4.5	Schwabe
1885[+]	S And	7	Hartwig, Gully
1887	V Per	9.0	Mrs Fleming
1891	T Aur	4.5	Anderson
1893	N Nor 1	7	Mrs Fleming
1895	RS Car 2	8	Mrs Fleming
1895[+]	Z Cen	7	Mrs Fleming
1898	N Sgr 1	4.7	Mrs Fleming
1899†	N Aql	7	Mrs Fleming

[+] supernova
[*] maximum later, in 1843
† discovered July 1900

followed by a slow decline. Nova Aurigae 1891 was discovered visually on 1 February 1892 by the Reverend T.D. Anderson at Edinburgh, at fifth magnitude. With this star the influence of photography came to the fore because examination of plates already taken on the field at Harvard showed that on 8 December 1891 the star was fainter than magnitude 13.2, but on 10 December it was of magnitude of 5.4. It reached its maximum of magnitude 4.2 on 17 December and then faded slowly, with fluctuations: an oscillating drop of 1.5 magnitudes in three months was followed by a drop of a further eight magnitudes in one month. Anderson also discovered the spectacular Nova Persei 1901 on 21 February of that year. Photographs revealed that the nova had risen from magnitude 12.8 to 2.7 in two days, reaching its maximum magnitude of 0.1 thirty-eight hours later. For eleven years it had been observed as a faint variable between magnitudes 11.0 and 13.0.

The first nova with any history prior to its outburst had been Nova T Coronae Borealis 1866, recorded in the *Bonner Durchmusterung* as of magnitude 9.5. J. Birmingham at Tuam in Ireland observed it as second magnitude on 12 May 1866, whereas it had been of less than naked eye visibility the night before. On 16 May William Huggins and W.A. Miller examined the spectrum and found that the light emanated from two sources. The principal spectrum was in absorption analogous to that of the Sun, while a second spectrum showed bright lines of hydrogen. Decades later, in 1946, this star was recognized as one of the type of recurrent novae. Another, T Pyxidis, was found by Miss Leavitt in 1913, for photographic records showed that it had been subject to outbursts in 1890 and 1902. And the puzzling spectra of novae, with their emission lines, did not merely aid their discovery, as when Mrs Fleming at the Harvard College Observatory discovered six novae between 1887 and 1900, but also introduced a major problem of interpretation.

The first ordinary nova whose spectrum was examined was Nova Cygni 1876: in September 1877, ten months after its maximum, R. Copeland at Dun Echt in Scotland found its spectrum to resemble closely that of an ordinary planetary nebula. When photographic spectra of novae were obtained they proved exceedingly complex and subject to great changes. Nova Aurigae 1891 was the first star of its kind to be investigated by 'the universal chemical method'. The spectrum was photographed in England on 3 February at Stonyhurst and South Kensington, and several days later at Harvard and Lick in the United States and at Potsdam and Hérény in continental Europe. On 22 February William and Margaret Huggins secured an exposure of $1\frac{3}{4}$ hours. The bright lines corresponded to the dark lines in the spectrum of Sirius, H and K were prominent, and there were many unidentified lines. Visually, the spectrum was very striking, with brilliant green rays and broad bright hydrogen lines, the red C line blazing "like a danger signal on a dark night" according to R. Espin. Later, the 'nebular lines' appeared in Nova Aurigae 1891, $\lambda\lambda 4363, 4959$ and 5007. They varied in brightness relative both to the hydrogen and to one another.

The first photograph of the spectrum of a nova after outburst but before maximum was of Nova Persei 1901. Miss Cannon is quoted as saying the Harvard workers were so surprised when it proved to be almost entirely an absorption spectrum that they thought the wrong star had been photographed.

A very comprehensive investigation of Nova Geminorum 1912 by A. Brill revealed that as a nova develops, the strong continuous spectrum with narrow absorption lines changes to weak continuous with broad bright bands. He showed the intensity of the continuous spectrum varied with the magnitude of the star and the bright emission bands did not reach their maximum until the star was well past its maximum.

The first pre-outburst spectrum obtained was that of Nova Aquilae 1918. After its discovery, a search revealed faint images on about ten Harvard photographs of spectra taken with the 8-inch (20-cm) telescope. Miss Cannon classified the best of these, taken on 1 July 1899, as having a nearly continuous spectrum with several dark lines in the hydrogen series, and with a distribution of light resembling Class B or A.

Identifications of spectral lines proved difficult because of the wavelength shifts, which astronomers assumed were due to the Doppler effect. In general, the shift was in proportion to the wavelength, as the Doppler effect required, but for some lines, the computed velocities were so large that observers wondered if some other effect was also

operating. The largest displacement was found for Nova Aquilae 1918, where the lines of ionized nitrogen at $\lambda\lambda 4097$ and 4104 indicated a velocity of approach of 3420 km per second.

Not until after 1900 was it firmly established that some novae have nebulosity around them. Earlier some observers had insisted that novae often had a fuzzy appearance compared with ordinary stars, but others asserted that this was merely the effect of uneven distribution of light in the spectrum. On 19 August 1901 C. Flammarion and E.M. Antoniadi photographed nebulosity about 6' in diameter around Nova Persei 1901. Later photographs showed an outward motion of this nebulosity of 11' a year, which J.C. Kapteyn suggested was the spread of illumination from the central disturbance. Knowing both the angular rate of expansion and the velocity of light, Kapteyn calculated the parallax of the nova as 0".011. Although fraught with ambiguities, the 'expansion method' would eventually provide the most satisfactory procedure for obtaining the distances, and hence the absolute magnitudes, of novae.

Much later, in December 1916, a different type of nebulosity was detected around Nova Persei 1901 when Barnard saw a faint ring with its centre close to the nucleus of the nova. By August 1919 the ring was about 10" in diameter with an outward motion of 0".4 per year. This was interpreted as the actual movement of matter out from the nova, and similar discs were found for Nova Aquilae 1918 and Nova Cygni 1920. In 1922 the Swedish astronomer K. Lundmark combined the observed radial velocity of expansion with the angular expansion of the gaseous ring around Nova Aquilae 1918 to derive a parallax of about 0".006. This was comparable to the average of several trigonometrical parallaxes for this nova, and because the various trigonometrical determinations exhibited considerable scatter, the expansion parallax competed favourably in acceptance with the trigonometrical results.

As the number of discoveries of novae increased, so early indications were confirmed, that novae are not scattered at random through the sky but are strongly concentrated to the galactic plane. Estimates could now be made of their frequency. From a systematic search of 1049 pairs of Harvard plates, Bailey in 1921 estimated that a series of bi-weekly photographs of the Milky Way with the

Harvard small-scale telescopes would yield nine novae per year. He concluded that the number of novae in the galaxy with apparent magnitude 9 or brighter would be twenty-five per year, one or two of which would be brighter than sixth magnitude.

From the time of the appearance of 'Tycho's nova' in 1572, the first of these great cataclysmic events recorded in the realm of the 'fixed' stars in early modern times, attempts had been made to offer a physical explanation. By the late nineteenth century astronomers were challenged to explain not only the cause of the star's outburst, but also the complex related phenomena, particularly the spectral variations, and the enormous velocities that occurred after maximum light. Isaac Newton had suggested a collision theory whereby a star, wasted by its emitted light, was reinvigorated by comets falling on it. A later theory, fostered by Laplace, considered the outburst to be a surface conflagration of a single star.

The opposed explanations, collision or explosion, were neatly brought together in the tidal theory of W. Klinkerfues. He saw the whole disturbance as coming from a single star, but as caused by tidal eruptions from the near approach of a second star (and he thereby laid the groundwork for modern theories of interaction in a double star system). In the period with which we are concerned, astronomical opinion remained divided between the descendants of Newton's and Laplace's theories. Each theory had many ramifications. For advocates of a collision the problem was to specify the nature of the colliding bodies. After it was shown that a faint star had existed in the position of Nova Persei 1901 before outburst, Norman Lockyer's suggestion of colliding streams of meteorites was ruled out. W.H. Pickering favoured a planetoid as the second body, while A.W. Bickerton considered a second star to be necessary. Among the collision theories, the most successful in accounting for the observed phenomena was that of H. Seeliger, in which the star encountered a dust cloud or gaseous nebula. Other theories involved the cooling of the crust of a dark star, or chemical combination at low temperatures, which caused a star to blow up and burst.

The wealth of observational material on the spectral changes of Nova Aquilae 1918 led both W.S. Adams and J. Evershed to conclude that the complicated spectra could best be explained as resulting from a shell of gas thrown off from the star at high velocities with both the forward and the backward edges rendered visible by huge Doppler shifts.

Novae in nebulae, Type Ib

The first indication that there might be a class of novae different from that previously recognized came with the outburst in the Andromeda Nebula in 1885. On the evening of 31 August of that year the Central Bureau in Kiel received at 10.38 p.m. a telegram sent from Dorpat at 10.15 p.m.: "Höchst merkwürdige Veranderung des grossen Andromedanebels fixsternartiger Kern siebenter Grösse. Hartwig." On 20 August Hartwig had first noticed the star that so greatly changed the appearance of the nebula, but he had withheld the announcement until possible spurious effects of bright moonlight had diminished. The first person actually to see the new star had been L. Gully at Rouen on 17 August, but he considered it might be an effect of his telescope optics. The nova, subsequently designated S Andromedae, increased from the ninth to the seventh magnitude by 31 August and then faded almost as rapidly as it rose, with one pause on the decline. Isaac Roberts obtained photographs on 3 and 5 September. The last sighting was on 7 February 1886 at about sixteenth magnitude. At least five observers studied the spectrum; some regarded it as continuous, but Huggins caught traces of bright lines on 2 September 1885, and Copeland measured bright bands on 30 September. In the annals of astronomy this star must be considered one of the most significant, but four decades elapsed before its significance was appreciated.

Ten years later, in another unresolved nebula, NGC 5253 in Centaurus, Mrs Fleming at Harvard found from its peculiar spectrum the nova subsequently named Z Centauri. Miss Cannon noted that the spectrum of Z Centauri resembled Class R and was unlike that of any other novae except that it bore a resemblance to that of S Andromedae. This lent support to the hypothesis that these novae actually belonged to the nebula surrounding each. (In 1936 Cecilia Payne-Gaposchkin on closer scrutiny noted that the spectra of Z Centauri and S Andromedae were not of Class R but resolved themselves into very wide bright lines.)

In 1909 a nova was discovered nearby the spiral

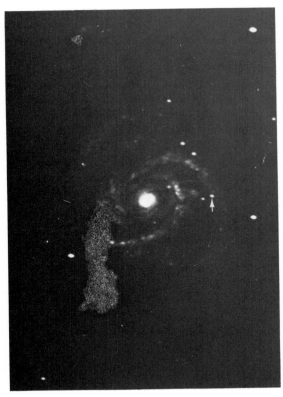 

5.3. Two novae in the spiral galaxy, NGC 4321, discovered in 1917 by H. D. Curtis on plates taken with the Crossley 60-inch reflector: a 3-hour exposure on 19 April 1901 (left); and a 2-hour exposure on 14 March 1914 (right).

nebula M 101 by Max Wolf of Heidelberg; but his announcement went almost unnoticed in North America, particularly after the star was classed as a long-period variable. Hence there was great excitement when G. W. Ritchey in the summer of 1917 discovered a nova of magnitude 14.6 in a photograph he had just taken of NGC 6946 in Cepheus with the 60-inch (152-cm) reflector at Mt Wilson. This nova was still visible, unlike the three which (it soon transpired) H. D. Curtis had already found on plates taken with the Crossley reflector at Lick. These announcements provoked a rush to study photographs of spirals taken with the great California telescopes, and a considerable crop of novae resulted.

Harlow Shapley (1885–1972), carrying on the suggestions of some late-nineteenth-century astronomers, computed the absolute magnitude of S Andromedae to be −16 *if* the nebula were an 'island universe', and he remarked that in his study of more than a million stars in globular clusters "not one is within ten magnitudes of this enormous brightness". But Curtis pointed out in 1920 that more than half the novae so far discovered in spirals were in the Andromeda Nebula, and that on average they were some ten magnitudes fainter than S Andromedae; the faintness of a *typical* nova in the nebula as compared with a typical galactic nova he ascribed to the enormous distance of the nebula. For Curtis, S Andromedae was exceptional, and "a division [of novae] into two magnitude classes is not impossible".

By 1924 E. P. Hubble had discovered novae and (more importantly) Cepheid variables on photographs of the Andromeda and other spiral nebulae taken with the new 100-inch (254-cm) reflector at Mt Wilson, and astronomers were soon agreed that such nebulae were indeed 'island universes'. S Andromedae was soon accepted as a representative 'supernova', as was Tycho's nova of 1572. A list of supernovae observed in nebulae is given in Table 5.2. It was, however, not until 1937 that two different groups of supernovae were recognized, a result of the systematic search programme of F.

Table 5.2. *Supernovae in extragalactic nebulae, observed before 1930*

Date announced	NGC	Constellation	Discoverer	Date of observed maximum	Magnitude of nova	Magnitude of nebula
1885	224	And	Hartwig	31 Aug. 1885	7.0	5+
1895	5253	Cen	Mrs Fleming	8 July 1895	7	10.8
1909	5457	UMa	Wolf	21 Feb. 1909	10.8	13.0
1917	6946	Cep	Ritchey	19 July 1917	14.6	11.1
1917	4321	Com	Curtis	17 Mar. 1901	13.5	10.8
1917	4321	Com	Curtis	2 Mar. 1914	14	10.8
1917	4527	Vir	Curtis	20 Mar. 1915	14	11.3
1917	2841	UMa	Pease	19 Feb. 1912	16	10.5
1920	2608	Cnc	Wolf	7 Feb. 1920	10.7	12.0
1922	4486	Vir	Balanowsky	24 Feb. 1919	11.5	10.7
1923	5236	Cen	Lampland	5 May 1923	14.0	10.0
1923	2535	Cnc	Reinmuth	18 Jan. 1901	13.3	<13.0
1925	4424	Vir	Wolf	17 Mar. 1895	10.5	12.6
1926	4303	Vir	Wolf & Reinmuth	9 May 1926	14.2	10.4

Revised from Helen B. Sawyer, *Journal of the Royal Astronomical Society of Canada*, vol. 32 (1938), 78.

Zwicky. A supernova that year in IC 4152, investigated by W. Baade and Zwicky, had a brightness which surpassed that of its entire galaxy by a factor of 100. After R. Minkowski had studied its spectrum, he reported to a meeting of the Astronomical Society of the Pacific in 1941 that nine supernovae formed a homogeneous group, which he designated as Type I, and five, distinctly different, he called Type II. The Type II spectra resembled normal novae, but he could find no satisfactory explanation for the spectra of the outstandingly luminous Type I, with broad emission bands.

Usual long-period variables, Type IIa

In 1800 only four long-period variables were known: Omicron Ceti, R Hydrae, Chi Cygni and R Leonis. During the nineteenth century a further 223 were found, 131 of them during the last decade – evidence of the efficiency of photography and spectroscopy. Of these variables only 15 are ever brighter than the sixth magnitude; but their usually large ranges, sometimes up to 9 or 10 magnitudes, made them an easy prey for sky watchers.

The Mira class, named for Mira (or Omicron) Ceti, came to be defined as those variables whose magnitude varies in periods of about 90 days up to more than 600 days (S Cas), with periods of 200 to 400 days the most frequent. Without exception the Mira variables are red, cool stars, and they are mainly of spectral class M, though they also occur in classes N, R, S and K. In the mid-nineteenth century Father Angelo Secchi described the spectrum of Mira as continuous with absorption; strong dark flutings terminated abruptly on the violet side and shaded off gradually in the other direction. The characteristic bright hydrogen lines were first detected in 1887 by photography at Harvard. In 1916, from a study of spectroscopic data, P.W. Merrill of the University of Michigan concluded that all longer-period variables show essentially similar spectra (then called Md) as described by Father Secchi, with absorption flutings of titanium oxide.

Long-period variables show considerable irregularity in their light fluctuations, height of maximum, or depth of minimum, and the cycle length may vary by as much as ten per cent. Accordingly, efforts to derive formulae to predict the light variations, which were so successful for eclipsing binaries, here proved very frustrating. Soon after the turn of the century, S.C. Chandler, H.H. Turner and T.E.R. Phillips derived elaborate formulae for the 610-day variable S Cassiopeae, from time to time adding more and more terms to account for the irregularities. In 1920 Leon Campbell at Harvard, who had made many thousands of observations of long-period variables, arranged seven classes of light curves in a logical progression of appearance. He developed the technique of smoothing out irregularities by taking 10-day means, instead of trying to represent them with elaborate formulae.

A host of theories were advanced to explain long-period variables. As early as the seventeenth century variables had been commonly taken to be rotating stars whose surfaces had dark regions; and after Alexander von Humboldt in 1851 made known to the world in his *Kosmos* the actual shape of the sunspot frequency curve that had been determined by S.H. Schwabe of Dessau, Rudolf Wolf was quick to note its similarity with the typical light curves of the Mira stars. He believed that the variation could be explained through periodic spot formations, either by differences in the observed surface brightness as the star rotated or by outbreaks of spots; but this hypothesis became untenable when it was realized that Mira stars are giants, as Henry Norris Russell (1877–1957) concluded in 1918. It seemed illogical to attribute to giant stars of very low density the surface phenomena of a smaller, denser star like the Sun. In the tidal theory of Klinkerfues, however, the variable was a close double in an eccentric orbit.

Explanation of the spectral changes was first attempted by A. Brester in 1908 with the suggestion that the light fluctuations were caused by the periodic appearance and disappearance of openings in a cool shell around the star. Merrill's studies led him in 1916 to propose an adaptation of this idea, the veil theory. According to this theory, at time of minimum a screen of condensed gas, perhaps calcium, exists in the upper atmosphere of the star, between it and the observer; the cloud formed by condensation conserves the heat radiated from the photosphere, and the temperature of the material just above the photosphere increases and the veil is vaporized. Such a phenomenon could be periodic and could also cause the observed variation in the spectrum.

U Geminorum stars, Type IIb
If the irregular behaviour of the long-period variables proved puzzling, the puzzle was minor compared to that posed by Type IIb, which spend much of their time at minimum brightness.

The prototype of the class, U Geminorum, was discovered by J.R. Hind on the evening of 15 December 1855 as a star of the ninth magnitude. By 10 January 1856 it was twelfth magnitude. Hind realized from its "very blue planetary light" and from his observations of minimum brightness extending over a great part of the period, that he had found a novel type of variable. The spectra of U Geminorum were found at Harvard in 1912 to be of Class F with broad H and K lines of calcium and faint hydrogen lines; the hydrogen lines varied in intensity and often appeared bright.

Other similar stars were gradually identified. The brightest of the U Geminorum stars was identified by Miss L. D. Wells at the Harvard College Observatory in 1896: SS Cygni, which varies from magnitude 7.2 to 11.2 in a period of about 40 days. In 1907 E. Silbernagel at Munich found a third and somewhat fainter U Geminorum variable, now called SS Aurigae, and by 1920 half a dozen other similar stars had been found. By 1922 the cause of the variation began to be understood when Adams and A. H. Joy showed the spectrum to have a close resemblance to that of novae, and indeed SS Cygni is now considered the brightest known member of a small group of stars known as the dwarf novae.

R Coronae Borealis stars, Type IIc
The behaviour of the R Coronae Borealis stars is opposite to that of the U Geminorum type: they spend most of their time at maximum light. In fact, they may remain at maximum for ten years before they suddenly drop many magnitudes to a minimum which itself may last for years.

Pigott discovered the prototype R Coronae Borealis in 1795 and it remains the best investigated. The record of its light curve from 1843 to 1905 was published at Potsdam in 1908 by H. Ludendorff and this work was continued by Campbell at Harvard. Normally the star stays at visual magnitude 6, but it may drop as faint as 15 at minimum. Spectra obtained as early as 1903 and 1905 by E.B. Frost at Yerkes and by Ludendorff showed great similarity to the luminous Alpha Persei. In 1922 Joy and M.L. Humason at Mt Wilson obtained spectra at minimum, and these showed bright emission titanium lines; at normal magnitudes the spectrum is cGO.

The other stars E. C. Pickering assigned to this class were RY Sagittarii, discovered in 1896 by E. E. Markwick, and SU Tauri, found by Miss Cannon and announced in 1908. By 1920 about a dozen stars were in this category, but some of these were later moved out.

Generally the variations of the R Coronae Borealis stars were explained as the effect of obscuration when a star moved through a cosmic cloud. Not until the 1930s did the abundance of carbon in these stars suggest the formation of a blanket of soot near the star.

Irregular variables, Type III
Variability of the type that Pickering called irregular showed up very early in the study of variables. Pickering's type star R Scuti was discovered as a variable by Pigott in 1796, and William Herschel noticed the variability of Alpha Herculis in 1795; Mu Cephei, which is sometimes taken as the type star of the class, was discovered by Hind in 1848. The irregular variables proved to be a miscellaneous group, and many were eventually moved to semi-regular classes. R Scuti, for example, was finally classed as an RV Tauri variable, a type of star in which deep and shallow minima alternate.

Short-period variables, Type IV
Pickering's Type IV comprised variables whose relationship to one another was puzzling. The common attribute was a relatively short period – a few days or even a fraction of a day – without the obvious binary characteristics of Type V. The name '*Blinksterne*' was applied to these stars well into the twentieth century, and the term 'Antalgol' was used for variables such as RR Lyrae whose brief maxima seemed the antithesis of the brief minima of Algol-type stars. In September 1784 Pigott discovered that Eta Aquilae was varying with a period of about 7 days, and that it increased in brightness more rapidly than it decreased. A few weeks later his friend John Goodricke found that Delta Cephei behaved similarly, with a period of between 5 and 6 days. In 1847 Argelander derived a 10-day period for Zeta Geminorum, a similar star which showed more gradual variations of light. By the end of the century the terms Delta Cepheids and Zeta Geminids came into use for variables of these kinds.

The light changes of these stars were quite different from those of eclipsing binaries. Nevertheless, the theory that the Cepheid variation might be due to eclipses in a binary system received an impetus in 1894 when, from 34 spectrograms with the 30-inch (76-cm) refractor at Pulkovo, A. A. Belopolsky discovered for Delta Cephei a velocity range from −19 to +24 km per second, and derived a velocity curve like that for a binary system. However, Belopolsky noted that the supposed time of minimum brightness occurred a day earlier than the supposed time of perihelion passage, and he wondered whether these two elements could be reconciled. Subsequently, the maximum velocity of approach was found to coincide with maximum light, and other changes, which did not support the binary hypothesis, were noticed in these variables: A. Wilkens in 1906 found the photographic range for them to be almost half again greater than the visual range, and during the following year S. Albrecht at Lick, from a light and velocity study of Y Ophiuchi and T Vulpeculae, showed that the peak of intensity of the spectrum shifted toward the violet as the light maximum approached, then back to the red as the brightness diminished. Nevertheless, as late as the 1920s, the elements of the orbits of Cepheid variables were published on the assumption that the light variation was due to eclipses in a binary system, generally with a highly eccentric orbit.

Some investigators sought alternatives to the binary hypothesis. As early as 1873 A. Ritter of Aachen had suggested that radial expansion and contraction in the star were the cause of the light changes. In 1909 F. R. Moulton noted that if a sphere is deformed slightly and left to its own gravitation it will oscillate around a spherical shape with a period that depends on the character of the oscillation and on the mass, density and elasticity of the star. Certain classes of variables, namely those such as Delta Cephei and Beta Lyrae, whose light changes constantly, may (he suggested) owe their variability at least partially to such non-radial oscillations.

In this same decade the appearance of the light changes of Bailey's 'cluster type' variables in the globular cluster Omega Centauri, even though the periods of these stars were all less than one day, indicated a possible relationship with the Delta Cephei class of stars. Furthermore, a star in a galactic field, RR Lyrae, which had been found by Mrs Fleming, had a similar curve, with a period of 0.5669 day and a range of 0.83 magnitude. In 1912 C. C. Kiess at the Lick Observatory, from a study of

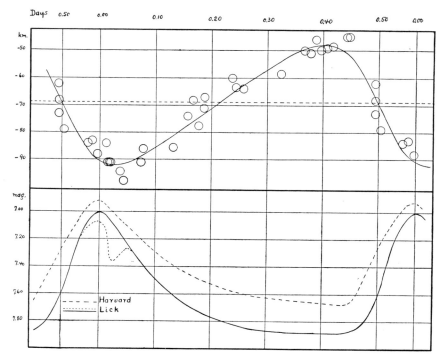

5.4. Velocity and light curves for RR Lyrae by C.C. Kiess, 1913.

both light and velocity curves of this type of variable, concluded that RR Lyrae and 15 other similar stars show the same type of variability as Bailey's cluster-type variables and can be considered to be the same class of star. Furthermore, he stated that these stars are part of the group of Delta Cephei and Zeta Geminorum stars and a separation into periods of under and over one day is artificial.

Meanwhile the various facets of Cepheid variation were gathered together by Shapley in 1914 in an historic paper in *The Astrophysical Journal*, "On the nature and cause of Cepheid variability". Shapley, too, reached the conclusion that the Cepheid and cluster variables are essentially the same. Using the discovery that Cepheids have great absolute magnitude (made independently by E. Hertzsprung and Russell), Shapley computed the density of an individual variable from Emden's polytropic gas sphere, and found exceedingly low densities for the redder spectral types. He ruled out the spectroscopic binary hypothesis because "these giant stars move in orbits whose apparent radii average less than one-tenth the radii of the stars themselves". He noted that a change from

one spectral type to the next would give a change of one visual magnitude and a colour change of 0.4 magnitude. Shapley concluded:

The explanation that appears to promise the simplest solution of most, if not all, of the Cepheid phenomena is founded on the rather vague conception of periodic pulsations in the masses of isolated stars. The vagueness of the hypothesis lies chiefly in our lack of knowledge of the internal structure of stellar bodies. . . .

Seizing upon this challenge, the English astronomer A.S. Eddington (1882–1944) undertook to develop the theory of Cepheid variation. Ultimately Eddington's work accounted for many Cepheid characteristics and also yielded an explanation of the empirical link between the period of a Cepheid and its absolute magnitude, the 'period–luminosity relation', to which we now turn.

After publishing her discovery of 1777 variables in the Magellanic Clouds in 1908, Miss Leavitt set to work to determine their periods. That year she gave the periods of 16 variables in the Small Cloud and noted that the brightest stars had the longer

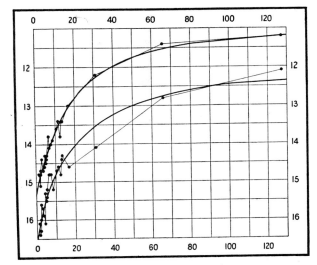 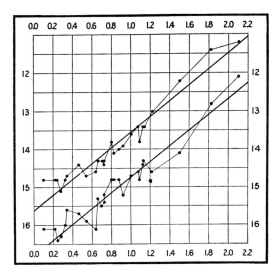

5.5. Period–luminosity diagram for 25 variables in the Small Magellanic Cloud by Henrietta S. Leavitt, 1912. Apparent photographic magnitude is plotted against the period in days (left) and against the logarithm of the period in days (right).

periods. In 1912 in Harvard *Circular* 173 she increased the number to 25 with periods ranging from 1.25 days up to 65.8 and even 129.7 days, the maximum magnitude ranging from 14.8 to 11.2. All but three of the stars had periods shorter than 17 days. Miss Leavitt commented that "the variables are probably at nearly the same distance from the Earth and their periods are apparently associated with their actual emission of light as determined by their mass, density and surface brightness".

Quick to recognize the importance of Miss Leavitt's discovery was Hertzsprung, who had been investigating the Delta Cephei stars and had shown that, because the stars have very small proper motions, they must be distant and luminous. In 1913 from a table of 13 such stars he concluded that on the average they are nearly one thousand times brighter than the Sun. Hertzsprung noted that, once the absolute brightness of a Cepheid variable has been derived from its period, its distance and position in space could be computed.

The master-touch to the period–luminosity relation was applied by Shapley. In 1917, through his work on globular clusters, Shapley had discovered a group of Cepheids that could be assessed in the same way as those in the Magellanic Clouds: the distance of any globular cluster from Earth is so

large compared with its linear diameter that for stars in the cluster an approximation to a constant difference between apparent and absolute magnitude may be assumed. Shapley found support for the period–luminosity relation from the analogous behaviour of a dozen variables in five globular clusters. His curve, with data combined from seven stellar systems, became known as 'Shapley's period–luminosity curve'. Shapley assigned a uniform median absolute magnitude of −0.23 to the RR Lyrae stars at the base of the curve. The new evidence, that the period–luminosity relation was valid in as many as seven different systems, suggested that it could become a yardstick for measuring the distance to *any* system for whose Cepheids the requisite data could be determined. (Not until after another three decades was Walter Baade to show that Cepheids in fact fall into two types, with a significant difference in their absolute magnitudes.)

While Shapley was pursuing his massive observational programme at Mt Wilson, Eddington was working in England to remedy "our lack of knowledge of the internal structure of stellar bodies". His classical paper on the transfer of radiant energy in stars (1916–17) was followed by another in the *Monthly Notices of the Royal Astronomical Society* of 1918–19, "On the pulsations of a gaseous star and the problem of the Cepheid variables". In this paper, Eddington used his theory of the 'standard' stellar model to derive the masses of 14 Cepheids from their luminosities. He next sought their radii and densities, but in the absence of detailed knowledge of the interior processes, he was unable to derive these parameters from the luminosities purely by theory. However, with an additional empirical parameter, namely the spectral type of each Cepheid, he could obtain their surface brightnesses. Combining this with known luminosities, he derived radii, mean densities and central densities. He found that for all Cepheids the period of variation, P, and the central densities, ρ_c, closely obeyed the relationship, $P\sqrt{\rho_c}$ = constant, as Shapley had previously shown in a qualitative way.

Eddington pointed out that the resulting radii were so large that the interpretation of a Cepheid as a binary required the companion to be inside the primary star, a conclusion he deemed "inadmissible".

Turning to the alternative 'pulsation' hypothesis, Eddington derived the theoretical values of $P\sqrt{\rho_c}$, which he showed to be nearly the same for all masses; hence it was a constant characteristic of his model. Moreover, the theoretical value was close to the empirical value, so Eddington realized that his pulsation theory would predict the observed period from the known luminosity and spectral type of a Cepheid.

The theory did not, however, explain why some stars pulsated and others did not, although Eddington was able to show that any pulsation, once initiated, would die out in a relatively short time unless there were an internal mechanism to sustain it. He suggested that this mechanism was related to a "critical state" of an evolving star, noting prophetically that "it is not unlikely that the specific heat may change with temperature, being abnormally high for temperatures at which ionization occurs rapidly".

Eddington made this suggestion more specific in his book, *The Internal Constitution of the Stars* (1926), where he proposed the existence of a 'heat value' that would cause the Cepheid to behave like an engine, converting the steady flow of thermal energy from the interior to mechanical energy in the outer layers.

Period changes were just beginning to draw interest in the interval under discussion. In his classical paper in 1918, already mentioned, Eddington computed the change of period necessary to "measure a very slight change in density, and so determine the rate of stellar evolution and the length of life of a star". Taking Delta Cephei as typical, he found that a period decrease of 40 seconds annually would be expected on the hypothesis that the heat of the star is provided by the energy of contraction. He considered that such a period change was too great to have escaped detection and that if Chandler's determined value of decrease of 0.05 second annually were correct it was evidence that the star's heat cannot be provided by contraction. The following year Hertzsprung took up Eddington's challenge for accurate determination of period changes for Cepheids. He showed from 18 existing series of observations that the period decrease of Delta Cephei was 0.079 ± 0.0083 seconds annually, thus confirming Chandler's value and Eddington's conclusion.

Variables of the Algol type, Type V

The suggestion that the light variability of a star could be caused by the eclipse of the star by a dark companion was first advanced in the eighteenth century. In 1782 Goodricke found that the variable Algol had a period of $2^d\,20\frac{3}{4}^h$ (a value he was later able to refine to an accuracy of seconds), and in consultation with Pigott, his friend and mentor, he offered the tentative explanation that a large planetary body was eclipsing the star. At the time his suggestion soon fell from favour because the variations in brightness seemed somewhat irregular, and because other known variable stars could not easily be explained in this way. But as the regularity of Algol's cycle was verified with the passage of the years, the eclipse explanation was revived, and by 1881 five stars were assigned to the class, all having a sharp drop to minimum light from an otherwise nearly-constant brightness. Furthermore, the eclipse hypothesis permitted computations of the actual properties of the two companion stars.

In 1881 Pickering reported to the American Academy of Arts and Sciences on the "Dimensions of the fixed stars with special reference to binaries and variables of the Algol type". He supported the eclipsing binary hypothesis, declaring that the regularity of the Algol variation precluded explanations that involved volcanic eruption, collision

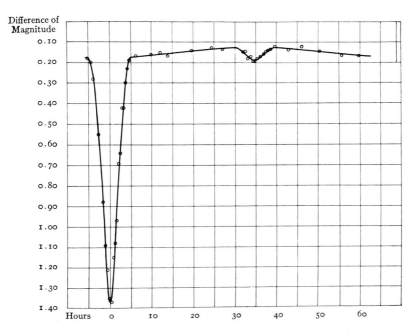

5.6. Joel Stebbins's photoelectric light curve for Algol, 1910. These results showed the secondary minimum for the first time.

or sunspots. About the same time, Pickering, C. Pritchard and P. Kempf introduced the first reliable and usable photometric measures of brightness for variables. Pickering gave equivalent diameters for stars of various magnitudes and, by assuming a diameter of 0″.006 for Algol, he derived four sets of elements for four different orbits.

The eclipse theory of the variability of Algol was splendidly confirmed in 1889 by H.C. Vogel with spectroscopic radial velocity measures at the Potsdam Observatory. The orbital elements for Algol were refined when for the first time the secondary minimum (when the faint companion is eclipsed by the brighter) was detected by Joel Stebbins with a selenium cell in 1910; Stebbins was then able to give the characteristics of Algol and its companion. A long-period oscillation in the radial velocity of Algol noticed by Belopolsky in 1906 and confirmed by R.H. Curtiss at Allegheny Observatory in 1909 and 1911 revealed the presence of a third body in the Algol system.

The possibility of computing dimensions and densities of eclipsing variables attracted great interest. In 1899 Russell published in *The Astrophysical Journal* computed mean densities for the 17 Algol stars then known. In terms of the density

of water, these ranged from 0.058 for Delta Librae to 0.728 for Z Herculis, with that of Algol at 0.139. Russell noted the relationship between the computed density and the accepted length of the eclipse. Vogel, assuming a different length of eclipse for Algol, derived a density twice as great; but Russell commented that the gradual change of brightness near the beginning and end of the light oscillation may be produced by causes other than the interposition of an opaque body. In 1905 James Jeans questioned the validity of the low densities, noting that the calculated density was so small that the original hypothesis of an incompressible fluid was quite untenable. If the densities were accurate, the system must be supposed to be in the gaseous state. He argued that knowing nothing but the period of revolution, we may not legitimately draw any inference as to the structure of the system.

Other investigators were at work on the problem. In 1903 A.W. Roberts of Lovedale, South Africa, had developed a method for determining the absolute dimensions of an Algol variable and J. Bauschinger of Berlin in 1906 formulated the relation between size, form and brightness of the bodies and the elements of elliptical orbits.

All the individual parameters were brought together by Russell in 1911–12 in his massive two-section paper in *The Astrophysical Journal*, "On the determination of the orbital elements of eclipsing binaries". His purpose, in which he was remarkably successful, was to develop formulae and tables to make the solution of eclipsing binaries simple. Russell noted that the stars of a close binary will appear, like the Sun, to be brighter at the centre than at the limb, and each star is brighter on the side that is heated up by its companion's radiation. Direct evidence of this was found by R. S. Dugan for RT Persei and Z Draconis and by Stebbins for Algol.

From Russell's groundwork, Shapley in 1915 developed his doctoral dissertation into a book-length *Contribution* from the Princeton University Observatory, entitled "A study of the orbits of eclipsing binaries". Shapley noted that "in October 1911 there were scarcely 10 eclipsing binaries for which the orbits were even approximately computed". With the Princeton Observatory polarizing photometer Shapley had obtained 10000 light measures and now, after two and a half years of observation and computation, he could derive the orbits of 90 eclipsing stars. He discussed such factors as darkening towards the limb, the relative dimension of components, ellipticity of stars, and the distribution of eclipsing systems with concentration to the plane of the Milky Way (as had already been noted by E. Zinner and P. Stroobant).

Different types of eclipses, partial or total, were recognized early. In 1903 Ludendorff determined the remarkably long period of 27 years for Epsilon Aurigae, confirming this value from Schmidt's 1874–75 observations of the star. This remains the longest period known for an eclipsing variable.

Several classes of eclipsing stars were gradually separated from the Algol type, depending upon the sizes, shapes, brightness and proximity of the stars. These distinctions include the W Ursae Majoris, Beta Cephei and Beta Lyrae stars.

By the turn of the century Beta Lyrae itself was recognized as an outstanding astrophysical problem. After Goodricke discovered its variability in 1784 and announced that it had two unequal minima, the star had continued to puzzle observers, as its complex light changes resembled no other. Changes in the spectrum, first noticed by Eugen von Gothard at the Hérény Astrophysical Observatory in 1882 and 1883, were found to be of a cyclical character; Hα, Hβ and then the D line of sodium were observed to be sometimes bright and sometimes not. In 1891 Pickering, using Mrs Fleming's observations, showed that two sets of radial velocities were involved. The following year Belopolsky at Pulkovo showed that the light changes and radial velocities might be explained by a system with partial eclipses. Almost five pages of summary in Müller and Hartwig's *Geschichte und Literatur* (1918–22) document the observations and attempted explanations for this star.

By 1914, 18 variables similar to Beta Lyrae were listed and Pickering assigned it Type Va of his classification. Only in later decades of the twentieth century did it become apparent that more than orbital revolution and tidal action are needed to explain complex binary systems; gaseous envelopes and streams of material passing between the two stars are essential in the model.

This ends the brief descriptions of the types of the 4400 or so variables discovered by 1920 and listed in the *Geschichte und Literatur*. These variables include 2007 in galactic fields, 47 novae, 570 in twenty globular clusters, a scant 16 in six open clusters, and 1791 in the Magellanic Clouds. From this wealth of material, significant discoveries of the content of the universe emerged. Variable stars fell into characteristic patterns, some of which were of major importance. The ordinary novae and the Cepheid variables in nebulae such as M 31 and M 33 provided the proof that these nebulae are indeed island universes. The very bright novae in nebulae led ultimately to the recognition that under rare circumstances a star could attain a luminosity comparable to an entire galaxy. The period–luminosity relation observed in Cepheid variables became one of the most important distance yardsticks available, and the investigation of the remarkable physical properties of these stars, proving that they were not binary systems, lent credence to this application. And intensive studies of light and velocity changes of eclipsing variables provided factual information about the dimensions and densities of stars that could scarcely be attained in any other way. Thus the light changes of the 'fixed' stars provided a far more comprehensive understanding both of the stars themselves and of the universe around them.

Further reading

Solon I. Bailey, *The History and Work of Harvard Observatory, 1839 to 1927* (New York and London, 1931)

Agnes M. Clerke, *Problems in Astrophysics* (London, 1903)

G. Eberhard, A. Kohlschütter and H. Ludendorff (eds.), *Handbuch der Astrophysik*, vol. 6: *Das Sternsystem, zweiter Teil* (Berlin, 1928), especially "Die veränderlichen Sterne" by H. Ludendorff (pp. 49–250); "Novae" by F.J.M. Stratton (pp. 251–98); and "Double and multiple stars" by F.C. Henroteau (pp. 299–468)

Caroline E. Furness, *An Introduction to the Study of Variable Stars* (Boston and New York, 1915)

M.A. Hoskin, Ritchey, Curtis and the discovery of novae in spiral nebulae, *Journal for the History of Astronomy*, vol. 7 (1976), 47–53

Cecilia Payne-Gaposchkin and Sergei Gaposchkin, *Variable Stars* (Cambridge, Mass., 1937)

Henry Norris Russell, Variable stars, *Science*, n.s., vol. 49 (1919), 127–39

W. Strohmeier, Variable stars, their discoverers and first compilers from 1006 to 1975, *Veröffentlichungen der Remeis-Sternwarte Bamberg*, no. 129 (1977)

6

Stellar evolution
and the origin of the Hertzsprung–Russell diagram

DAVID DEVORKIN

Gravitational energy

How are the Sun and the stars apparently able to give out light and heat in great quantities without themselves wasting away in the process? This question was interesting but not crucial until the mid-nineteenth century when it was recognized that the energy store of the Sun, based upon every known source of energy, was grossly insufficient to have allowed it to shine for the eons required by uniformitarian geology. Even before this recognition, the founders of the new science of thermodynamics realized that they could apply the concept of the mechanical equivalence of heat to the question of the source of solar energy. In doing so, R. Mayer, J. Joule, W. Thomson (later Lord Kelvin), H. v. Helmholtz and J. Waterson established for the first time rational theories for the source of energy of the Sun and the stars, based upon the conversion of gravitational potential energy into heat and light. Except for Helmholtz, their first theories were mechanical inventions based upon meteoritic collisions. Since a bright meteor passing through our atmosphere can achieve a brilliancy that rivals the Moon when full, meteoritic bombardment on a vast scale was seen as the way to convert mechanical energy into heat at the surface of the Sun. Similarly, the original store of solar and stellar heat was imagined to exist in a vastly extended cosmic cloud composed of meteoroids. Their mutual gravitational attraction would cause the cloud to collapse, and through collisions the kinetic energy of motion of the bodies would be transformed into light and heat. A star could thus be born as a hot body from a cold, invisible, nebulous cloud.

This theoretical mechanism soon proved to be inadequate, for by 1860 dynamical studies of the orbits of the inner planets showed that the maximum allowable amount of infalling meteoritic material was insufficient to sustain the Sun's luminosity. Furthermore, if the Sun's luminosity were presently maintained by meteoritic bombardment, the Earth would be bombarded at a rate sufficient to make our own atmosphere incandescent. Nevertheless, Thomson, who was the only one to carry this idea through, did not abandon meteoritic bombardment. Instead, he modified his theory by placing the meteoritic phase of the Sun's heat in the distant past; the present Sun was a convective mass of liquid and vapour continuing to contract and cool. This revised theory, now similar to that of Helmholtz, still could not account for the enormous time scales required by uniformitarian geology. Even so, Thomson's suggestion that stars existed in convective equilibrium and his argument that the sole source of energy of the Sun was gravitation remained almost unquestioned by astronomers for over fifty years. Thomson argued vigorously against the time scales required by uniformitarians, and in doing so kept open the question of vast time scales. It is within the context of gravitational contraction, then, that we now review late nineteenth- and early twentieth-century progress towards an understanding of the Sun and stars as physical bodies.

Studies of solar and stellar interiors

The first indications that the Sun was not a solid came in 1859. In that year, the English amateur R.C. Carrington showed solar rotation to be variable with solar latitude – impossible for a solid body – and G.R. Kirchhoff established the gaseous nature of the solar atmosphere. In the 1860s, laboratory evidence supported the contention that the interior of the Sun was gaseous; the Irish physical chemist Thomas Andrews showed that for all gases, there was a 'critical' temperature above which no amount of pressure could cause liquefaction, and so the solar interior could remain in a gaseous condition even though great pressures were certain to exist there. Finally, J. Plücker, E. Frankland and F. H. A. A. Wüllner all showed that a gaseous discrete line spectrum could be turned into a continuous spectrum (that produced by the solar interior) if sufficient pressure were applied.

H. Faye (1814–1902), in the mid-1860s, was the first to exploit extensively these new developments and to discuss the Sun as a gaseous sphere. Faye's work prompted the American J. H. Lane (1819–80) in 1869 to re-examine some earlier calculations he had done during the American Civil War on the physical structure of the Sun, which were designed to investigate if the Sun behaved as a perfect gas sphere, given its observed surface conditions. Now realizing that his original theory was inadequate, Lane reversed his approach; assuming that the Sun behaved as a perfect gas, he examined its theoretical temperature distribution and attempted to reproduce its observed surface conditions. He was not fully successful since his derived surface temperature of the Sun could not agree with observed estimates unless somehow the known gas laws did not really conform to the kinetic theory of gases – something neither Lane nor anyone else was willing to consider seriously. Further, although Lane made no great point of it, estimates for the observed surface temperature of the Sun differed widely, making any comparison with theory inconclusive.

Lane, an otherwise obscure member of mid-nineteenth-century scientific Washington, was nevertheless the first to construct a theoretical model of the Sun wherein one could determine the temperature, density and pressure found at any point within the solar interior. His unpublished notes in the US National Archives show that he also considered what could happen to the surface temperature of the Sun if its radius were to decrease through gravitational contraction. Lane found that as long as a star maintained the perfect gas condition (that is, as long as it behaved according to the gas laws of Boyle and Charles) any contraction of its surface would cause an homologous rise in temperature, including the surface temperature. If the star achieved a density at which it no longer behaved as a perfect gas, further contraction could be accomplished only through cooling.

In the late 1870s, A. Ritter (1826–1908), Professor of Mechanics at the Polytechnical School in Aachen, examined the increase in temperature that would accompany the contraction of a perfect gas sphere, apparently unaware of Lane's unpublished work. In eighteen papers in a span of six years, he provided a general discussion of the behaviour of a gas sphere in convective equilibrium under its own gravitation, examined what the future of the Sun might be, and created a theory of stellar evolution based upon convective equilibrium. Ritter's theory differed from the conventional view of stellar evolution as a continual cooling process by including an initial heating phase that occurred while the star behaved as a perfect gas, in accordance with what both he and Lane had theoretically derived. In so doing, Ritter was the true founder of what we will refer to as the 'Giant and Dwarf' theory of stellar evolution, later to be advanced by Norman Lockyer and Henry Norris Russell.

Ritter's work remained unappreciated until the 1890s when it was brought into discussions of stellar evolution by William Huggins in England; at Huggins's urging, George Ellery Hale in America had one of Ritter's papers translated and reprinted in the new *Astrophysical Journal* in 1898.

During this period Thomson continued to discuss his own highly revised theories, but with Ritter and others could not decide exactly how they were to be applied: there was simply no way to determine from available observational evidence whether the Sun and stars behaved as perfect gases, or, if they did, when they would cease to do so. Even though a handful of astronomers were deriving methods at that time to determine the densities of stars in binary systems – density being precisely the datum needed to decide if a star behaved as a perfect gas – the general opinion was that an understanding of the applicability of the gas laws to stars was not yet at hand.

In 1907, R. Emden (1862–1940), an assistant professor of physics and meteorology at the Technische Hochschule in Munich, published a treatise called *Gaskugeln*. This was a textbook for instruction at the Technische Hochschule and was designed to attract students to an abstract area of physics by employing, as 'practical examples', calculations of the structure of stars. The book was a masterful synthesis of the work of Lane, Thomson, Ritter and others on the problem of convective equilibrium, and its appearance did much to stimulate interest in the field – not least because Emden provided extensive tables of functions required for the general solutions of the equations of convective equilibrium, which simplified the process of computation.

After treating the problem of convective gas spheres, Emden considered in turn models of stars with rigid crusts and/or centres, the solar limb darkening, the Earth's atmosphere, and pulsating gas spheres. One of Emden's most significant discussions concerned the theoretical conditions necessary for a gas sphere to have a boundary with finite radius (where density and pressure go to zero) and hence to behave like a real 'star'. In a later part of his book he attempted to apply his mathematical techniques to the construction of models for specific stars, and introduced an added degree of sophistication in the form of polytropic models that allowed for any degree of central condensation within the star.

Although Emden's work is recalled today as the culmination of the early era wherein stars were treated as perfectly convective spheres of gas, contemporary commentators, notably James Jeans, were conservative in accepting the applicability of thermodynamics to the stars. Jeans in particular argued (in a review of *Gaskugeln* in *The Astrophysical Journal* in 1909) that too little was known about the physics within the stars to justify modelling them strictly according to the known laws of thermodynamics, disregarding the still largely unknown effects of radiation pressure, electrical action and other agents. But as the physics within the stars became better known in subsequent decades, much of Emden's work survived and was adapted, along with that of Ritter, to studies of stellar structure.

Stellar evolution and spectral classification

To appreciate how the physics of the stars became better understood, we must examine the origins and development of the major systems of spectral classification of stars, and their associated theories of stellar evolution. Two main theories emerged: one, that stars began as hot blue stars, and

contracted and cooled to become red stars; the other, due to Ritter, that stars began as vast cool red stars, heated upon contraction to become blue stars, and then cooled to become red stars again. While the first theory remained dominant until the work of Russell, elaborations of it in the late nineteenth century did include the introduction of preliminary heating phases. But the main problem that persisted was whether such a preliminary heating phase was manifest in the observed sequences of spectra among stars. Lockyer argued that it was, and almost everyone else argued that it was not.

J.K.F. Zöllner (1834–82), a highly influential pioneer astrophysicist based at the university in Leipzig and creator of many early photometric and spectroscopic devices, was one of the first to examine systematically the luminosities of stars and to apply his findings to the development of an evolutionary sequence. In his pioneer work *Photometrische Untersuchungen* (1865), he attempted to link the observed spectral sequence of stars and the progression of colour manifest in this sequence to an evolutionary cooling sequence. He suggested five stages, starting with annular (or planetary) nebulae composed of luminous gases (this being suggested by the work of Huggins, who had shown the previous year that some nebulae exhibit bright-line spectra characteristic of rarified gases). Subsequent cooling caused the luminous gas to liquefy: with the formation of a solid crust, which would crack and buckle releasing hotter liquids from time to time, the star became variable in brightness. Finally, the crust thickened to the point where it could contain its turbulent interior, and the star cooled into extinction.

Unlike Herbert Spencer in his speculations in 1858, or Zöllner's French contemporary Faye, Zöllner compared his evolutionary model with observations, and in so doing influenced his student H.C. Vogel (1841–1907) to continue the work utilizing the spectra of stars. In the 1870s, Vogel, who had been conducting spectroscopic research at Bothkamp (near Kiel) and was soon to become one of the first staff members of the new Potsdam Astrophysical Observatory near Berlin, created a system of spectral classification based upon the idea that stars cooled with age. Even though the causes for differences in the spectra of stars of different colours were not understood at the time, the fact that the stars were classifiable into so few major groupings indicated that some consistent physical mechanism was at work rather than the random influence of differences in chemical composition. This view was first advocated by the Jesuit astronomer Father Angelo Secchi (1818–78) in the 1860s in his identification of four types of stellar spectra, and was shared by most classifiers in the late nineteenth century. But Vogel, possibly the most influential of them, did allow for a variation in composition in his scheme, and in 1874, with only crude visual spectra and a small sample, he proposed a linear system in which three classes of stars were distinguished by colour: white, yellow and red. Though Vogel's system was similar to Secchi's it differed in combining Secchi's Types III and IV (both red classes) into IIIa and IIIb. Throughout the 1880s and 1890s, with improved spectral criteria available partly through the application of spectrum photography, Vogel added subclasses. The order of his system, however, remained essentially unaltered.

Norman Lockyer (1836–1920) was the founder and editor of *Nature* and a pioneer solar and stellar astrophysicist. Like his contemporary Huggins, he was a brilliant and intense Victorian amateur-turned-professional based in London. Lockyer objected to Vogel's combination of the red stars into one class with two subdivisions; for reasons to which we come shortly, Lockyer wished to place some red stars into these subdivisions at the beginning of the evolutionary sequence and some at the end. Lockyer's interest in the physical and chemical nature of the Sun's surface began in the 1860s, and by 1873 extended to the spectra of stars, in this relying upon the observations of others. Lockyer himself had conducted extensive laboratory investigations of spectra, and these suggested to him that differences seen in celestial spectra were due to temperature differences in stellar atmospheres. He was the first to propose that, with increased temperature, atoms in stellar atmospheres would break down into their fundamental constituents. This was his concept of 'celestial dissociation', an extension of his laboratory observations of spectral changes brought about by increased temperature. Compounds could exist in the cool red stars, while in hotter yellow stars, compounds were broken down into atoms. In the hottest blue stars, the atoms themselves were broken down into the fundamental building blocks of matter. The nature of these building blocks was far from clear, and very few could follow Lockyer's reasoning because most physicists were strongly tied to the concept 'one element – one spectrum' and to the belief that atoms were indestructible.

Undaunted, Lockyer in 1873 applied his ideas to a tentative linear classification scheme and then in the 1880s he altered this evolutionary sequence into his famous 'temperature arch' based upon his so-called 'Meteoritic Hypothesis'. In 1887 Lockyer concluded from his laboratory findings that the spectra of nebulae and of vaporized meteoritic material were the same, and therefore that nebulae were composed of swarms of meteoritic material, "the collisions of which bring about a rise in temperature sufficient to render luminous one of their chief constituents – magnesium". Within the next year he developed his full theory of stellar evolution based on the concept of dissociation and the meteoritic interpretation of nebulae.

His basic notion was that stars began as vast extended swarms of meteoritic material contracting under the influence of gravitation. As contraction proceeded in the cloud, the temperature would rise causing the spectra of the volatilized meteoritic material to change. The changes observed among the line strengths of different compounds and elements were interpreted as due to the rise of temperature and to the varying volume occupied by the stones, heated gas and regions of collision. Contraction would continue, with subsequent heating causing the colour and spectrum of the object to vary from red to blue. Finally, the entire mass would volatilize and the object would become truly stellar. But now, no longer heated by collisions of meteoritic material, the star would cool from blue to red and hence pass to extinction.

Lockyer never wished to link his temperature arch to the physical theory of Ritter and Lane, for he, like most astronomers, could not convince himself that criteria were available for deciding when a star behaved as a perfect gas and when it did not. Furthermore, Ritter believed that only the most massive stars would attain the highest temperature and that though all stars would pass through the red stage twice, the actual temperature path would depend upon mass. Lockyer, however, adhering to strict uniformitarian meth-

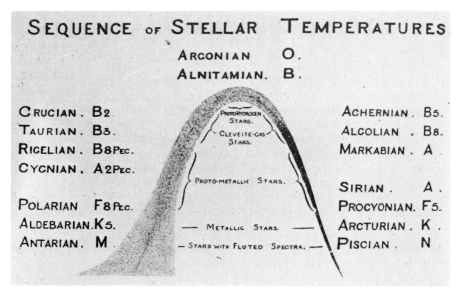

6.1. Norman Lockyer's 'Temperature Arch' of 1914, embodying his theory of dissociation and the Meteoritic Hypothesis. Stars begin their lives as extended clouds of meteoritic stone at the base of the left-hand branch; as the cloud collapses, meteoritic collisions cause the temperature to rise until all the solid material is vaporized. Subsequent contraction is accompanied by cooling, bringing the star to extinction at the right-hand base of the arch. Included along the branches are examples of Lockyer's classification in the 1890s (his archetypal genera) and their Draper classification equivalents.

odology and guided by his theory of dissociation, required that all stars experience the same evolutionary course.

Lockyer's use of the meteoritic mechanism was not unique, however. He was influenced by P. G. Tait, who in turn was a close associate of Thomson who believed in a meteoritic bombardment phase in the early history of the Sun. At about the same time the geologist J. Croll outlined a cosmogony based on collisions of large bodies, and George Darwin, son of the great biologist Charles Darwin, endeavoured in the late 1880s to reconcile Lockyer's meteoritic hypothesis with the Nebular Hypothesis.

Despite Lockyer's efforts the dominant concept of the direction of the evolution of stars as revealed by stellar spectra remained a linear progression of cooling from blue to red. But it is important to note that many classifiers, including Vogel and Huggins, recognized the need for some heating phase in the life of a star: a hot star must somehow evolve out of a cold nebula; since Huggins's 1864 discovery of the bright-line spectra of nebulae, the older 'shining fluid' concept of nebulae had been

replaced by the belief that nebulae were, indeed, quite cold. Vogel believed that the amount of time a star spent in this heating phase was short compared to stellar lifetimes, and so he concluded that no stars had been observed in this phase. Huggins, in the 1890s, was sympathetic to the Lane–Ritter theory but suggested that most bright stars were actually heating, even though their colours progressed from blue to red with age. To Huggins, this colour progression was illusory, caused by increased atmospheric masking with continued condensation, and stars cooling to extinction were too faint to be observed. Lockyer, then, was the only classifier of spectra who maintained that both the heating and the cooling phases of stars were observable among the stars in the sky bright enough to be classified by their spectra.

Throughout the late nineteenth century, one of the main difficulties in reaching a consensus about the direction of evolution was that no clear understanding existed of the causes for the variations seen in stellar spectra. Line width, differences in line structure, and the presence and absence of lines, were thought to be manifestations of abundance differences as well as differences in temperature, pressure, density, and, in Huggins's opinion, atmospheric masking. Definite temperature criteria began to appear only with the careful work of Lockyer's long-time associate A. Fowler (1868–1940) in London and of Hale at Yerkes Observatory. They independently showed that the spectra of red stars resembled sunspot spectra. Through comparison with laboratory spectra and other evidence they also knew that sunspots were cooler than the general surface of the Sun, which was classed as a yellow star. Thus, broad temperature/colour criteria began to be established, but with the proliferation of spectral classification systems (twenty-three systems were in existence by 1900) there was little hope that a useful universal system could be found.

In 1904, E.B. Frost (1866–1935) at Yerkes attempted to surmount the confusion by inducing astronomers to come to a consensus about classification. Frost felt that progress would be made on understanding stellar evolution only when classification had been made both systematic and physically comprehensible. But there was much to be clarified in observational astronomy before any consensus could have meaning. Specifically, what was the physical basis of spectra? Was there any relationship between the spectrum of a star and its intrinsic brightness or location in the sidereal system? Were stars all of the same size? At the time, these questions could not be answered. Reliable data on stellar distances and brightnesses were still lacking, and astronomers were uncertain whether the new studies of radiation by physicists such as Josef Stefan, Wilhelm Wien and Max Planck could be applied to the stars.

Reliable measurements of the distances to stars, and hence of their true luminosities, were possible only by the method of trigonometric parallaxes. By the 1860s, with the use of visual techniques, barely a dozen stars had measured parallaxes that significantly exceeded their probable errors. By 1900 this number had risen to about a hundred, although proper motions had been determined for many more stars. Statistical analyses that would provide relative distances of various spectral groups were just beginning. Improved methods of determining relative apparent brightnesses of stars were yielding large bodies of data, as were the great spectral classification and photometric projects at

Harvard. But the photometric projects were not consistent from one observatory to another, and they contained systematic errors preventing them from being reduced to a physical system for the measurement of radiation. In Potsdam, however, the staff, headed by Vogel and J. Scheiner, were actively engaged in photometric surveys that were carefully calibrated to an absolute system of intensity measurement; but these were just starting at the turn of the century.

The Harvard spectral classification and the origins of the Hertzsprung–Russell diagram

Probably the most important step in the development of a comprehensive system of stellar spectrum classification was the "Draper catalogue of stellar spectra" published in various volumes of the Harvard *Annals* in the 1890s. In the mid-1880s, E.C. Pickering (1846–1919), the director of the Harvard College Observatory, had initiated and maintained two vast projects to acquire reliable and consistent spectroscopic and photometric data for stars. For both projects, partly funded by the family of Henry Draper, Pickering provided innovative techniques that allowed for consistent work on a large scale. In his spectroscopic project, he employed wide-field photographic astrographs equipped with thin objective prisms so that hundreds of stars could be photographed for their spectra at once. These plates could then be examined in the relative comfort of an office, and a large staff were kept occupied studying many plates at once for the identification of the stars and the assignment of spectrum class. In this manner, Pickering's team, led in the early years by his chief assistant, Williamina P. Fleming (1857–1911), brought out in volume 27 (1890) of the Harvard *Annals* the first Draper catalogue containing spectra for over ten thousand stars. For this catalogue Pickering created an alphabetical classification based upon the Balmer series and the H and K lines of calcium. Line continuity was his major criterion; hydrogen was dominant in Class A, and decreased in strength through B, C, D, becoming invisible well before his ultimate Class Q. This series was based upon Secchi's four types: Type I contained Classes A to D; Type II contained E to L; Type III was represented by Class M; and Type IV by Class N. Classes O, P and Q contained bright lines and did not resemble any of the other classes.

The first Draper catalogue not only doubled the number of stars classified by spectrum but allowed for the first time the pursuit of statistical studies based upon spectra. Pickering himself was interested in the spatial distribution of spectra and actively pursued its study. In like manner, soon after the appearance of the first catalogue, two men independently compared large samples of proper motion data of stars selected by spectral type in order to determine relative distances to these groups of stars: W. H. S. Monck (1839–1915), a brilliant amateur astronomer in Dublin, and J. C. Kapteyn (1851–1922) of Groningen, revered today as the originator of statistical astronomy and the modern study of galactic structure. The relative distances, combined with the apparent brightnesses of the stars in each group, permitted the determination of the ratio of their mean absolute luminosities.

By 1892 Monck had found a strange correlation: as he examined the Draper spectral types in turn, from the blue stars (A and B spectral classes) to the yellow stars (F and G), the proper motions increased steadily and then decreased, so that the yellow stars had the largest values rather than the red ones. Later that year Kapteyn found the same relation.

The result was perplexing for two reasons. First, since proper motions are statistically an inverse function of distance, the group with the largest motions must be statistically the closest to the Sun; that is, the yellow stars, as a class, were closest. And second, when their average observed apparent brightness was compared to that of the red and blue stars, the yellow stars proved to be the least luminous.

The first finding – that the yellow stars were closest – was simple to reconcile with the idea that the Sun was part of a small cloud of yellow stars. Though this idea seemed reasonable, it again raised the possibility of abundance differences rather than changes in physical condition (temperature, pressure) as the causes of differences in spectra. The more intriguing, and unexpected, result was the second one – that yellow stars were less luminous than the red or blue stars. This caused trouble for the commonly accepted theory that evolution involved contraction and cooling

from blue to red. For the accepted theory to be correct, yellow stars must be bigger than red stars; but it now seemed that the yellow stars were smaller than the red stars by reason of their lesser luminosity, and by 1893 both Kapteyn and Monck had come to this conclusion. Monck then proposed that the colour of a star evolving by contraction passed through blue, then red and then yellow. Kapteyn, however, made no attempt at the time to relate his work to stellar evolution.

After the turn of the century, largely through Kapteyn's efforts, more extensive and reliable data on proper motions became available. But in addition to collecting data, Kapteyn created a refined statistical method for measuring stellar distances, the technique of 'secular parallaxes'. Using proper motions and radial velocities to determine relative space velocities, Kapteyn found a clear increase in velocity from the B stars to the early M stars of the Draper system. The 'Helium stars' or B stars had the lowest linear space velocities, comparable to those of diffuse nebulae like the Orion Nebula, and these Kapteyn believed to be the youngest stars. The planetary nebulae had the highest velocities of all, and from this Kapteyn concluded that planetary nebulae could not produce stars. Instead, he was willing to regard planetary nebulae as the *final* stage of stellar evolution, because on at least two occasions in the previous twenty years, stars that had become novae had been seen (spectroscopically by W. W. Campbell and photographically by G. W. Ritchey) to change into nebulae.

Neither Monck nor Kapteyn was able to understand the peculiar low luminosities of yellow stars, although Monck offered some speculations on the matter. In 1894, in a paper in the *Journal of the British Astronomical Association* entitled "The spectra and colours of stars", he concluded that there were probably two distinct classes of yellow stars, "one being dull and near us and the other bright and remote" like the blue stars. To Monck, Capella was an example of a distant yellow star many times more luminous than the Sun, while Alpha Centauri was a very close yellow star with a luminosity similar to the solar value.

Monck's interesting statement, certainly a clue to the existence of giant and dwarf stars, was not developed further except in some rough calculations by J. E. Gore (1845–1910), who showed that the red star Arcturus was a 'giant star' because its diameter was about eighty times that of the Sun, or about the size of the orbit of Venus. Gore used W. L. Elkin's Yale heliometer measurements of the parallax of Arcturus, which came to 1/60 second of arc, to derive a distance of twelve million astronomical units (the Sun's mean distance from Earth). Using the inverse square law of light, Gore calculated that this distance ratio meant that the Sun would appear over 35 magnitudes fainter if it were removed to the distance of Arcturus, fully 9.5 magnitudes fainter than Arcturus itself. Thus Arcturus had to be 6300 times brighter than the Sun, and, on the assumption that the surface brightness of Arcturus was the same as that of the Sun, Arcturus had to be 79 times larger. Gore did not provide exact values for his estimates, on the grounds that the measured parallax was too small to be reliable; but he believed that his general result was meaningful, though contemplation of it would "stagger the imagination". Neither Gore nor Monck mentioned the possibility that giant red stars would fit quite well into the theories of Ritter and Lockyer.

In the 1890s, to resolve the apparently anomalous behaviour of the yellow stars, Monck and Kapteyn needed better data, both astrometric and photometric. Monck's speculation might have been bolder if he had known that Antonia C. Maury (1866–1952) of the Harvard College Observatory, one of the women working for Pickering on spectra, was finding that within each of the bluer spectral classes, two distinct types of spectra existed. Until the arrival of Miss Cannon in 1896, Miss Maury was the only woman on the Harvard staff trained in astronomy and physics. She was the niece of Henry Draper, and so was acutely aware of the importance of spectral classification in the growth of astrophysics. Pickering appropriately assigned Miss Maury to a unique task within his group.

The primary goal at Harvard was to compile as large a list of stellar spectra as possible, with the sacrifice of detailed knowledge of the spectra of each star. But Miss Maury was asked to examine a small set of stars in greater detail. In 1897, after nearly a decade of preparation, she published her findings, and proposed a new classification system consisting of twenty-two groups. Within each group she identified subdivisions based upon the characteristics of the spectral lines. Two main

subdivisions were constructed for the bluer stars: stars with normal line intensities where hydrogen lines were broad and dominant while metal lines were weak and diffuse, and stars with abnormal line intensities where the hydrogen lines were narrow and well defined and the metallic lines sharp and stronger than in the normal stars. The designation 'normal' was based simply on the frequency of occurrence of the spectra. The abnormal cases were relatively rare; these were designated as spectral subdivision 'c'. We will refer to them as c stars, and to the normal spectra as non-c stars.

Miss Maury's classification differed from the alphabetical Draper classification supported by Pickering. Miss Maury also associated her c and non-c stars with evolution, speculating that they represented "parallel courses of development". There is no evidence that Miss Maury knew why the spectra differed, or that she was aware of Monck's work, or he of hers. Indeed, her work went unappreciated at the time and remained so until E. Hertzsprung (1873–1967), a young Danish astronomer trained in photochemistry, decided to determine just why the two subdivisions existed, and what caused them to differ.

Hertzsprung stated at the outset of his paper, "Zur Strahlung der Sterne I", which appeared in *Zeitschrift für wissenschaftliche Photographie* in 1905, that it was Miss Maury's classification that had stimulated him to take a look at the problem. But he was also influenced by his background in photochemistry and by his familiarity with Planck's recently announced radiation law. He therefore knew that from the radiation characteristics of a star – its apparent brightness and colour – one could, by using Planck's law, determine the angular size of the source; and with further knowledge of the distance of the source, determine the actual linear diameter. In 1906 Hertzsprung used this approach to discuss the diameter of one star, Arcturus, and this he found comparable to the diameter of the orbit of Mars. Thus Hertzsprung knew of the existence of very large stars, and his familiarity with the work of Kapteyn and Monck is shown by the references in his statistical papers published in 1905 and 1907. It is therefore possible that he suspected that what Miss Maury had detected in her spectroscopic studies could be explained by a careful examination of the peculiar luminosity distribution found by Kapteyn and Monck.

Following methods closely related to Kapteyn's, Hertzsprung first reproduced the distribution found by Kapteyn and Monck and then went on to separate out the c and non-c stars, examining them independently as a function of apparent brightness and proper motion. He quickly found that the c and non-c stars had different luminosities, and that while the c stars were very distant (indicated by their small proper motions), they were also among the brightest in the sky, so the range of luminosity was vast. Though Hertzsprung had established this result by use of proper motions, he then verified it with parallaxes that were available, and these confirmed his findings completely.

Hertzsprung's work explained why Kapteyn and Monck found that the group with the largest proper motion were the faint yellow stars and not the red stars. Without the c characteristic as a discriminant, high and low luminosity red stars would be grouped together, as would high and low luminosity yellow stars. According to Hertzsprung, the luminosity range of the red stars was considerably greater than for the yellow stars, so that the less luminous red stars were decidedly fainter than the least luminous yellow stars. Thus observational selection worked against having enough faint red stars in the sample, and the group was biased in favour of the distant, highly luminous red stars, which of course showed little proper motion. Hence the mixing of bright red stars caused the mean proper motion of the whole group to be less than the mean for the mixed yellow stars.

During 1906, Hertzsprung continued to examine and refine his discoveries, and in 1907 he published another paper on the subject clarifying the effects of mixing and discussing the fact that the highly luminous c stars were very rare. Through all this period Hertzsprung worked in Copenhagen as a 'private' astronomer enjoying the use of several local observatories, and it was there that he began his lifework on the photographic photometry of star clusters.

After the publication of his first papers, Hertzsprung's local reputation grew. He became a correspondent of the influential German astronomer K. Schwarzschild (1873–1916) of Göttingen; and in addition he corresponded with Pickering – to whom he sent his initial findings – and with Frost of the Yerkes Observatory, whom he asked for data. Thus one would expect Hertzsprung's work to have become generally known, by 1907 at least; but in fact such was not the case.

Meanwhile, at Harvard, Pickering was pushing ahead with a refinement of the Draper alphabetic spectral classification system. This was largely the work of Annie J. Cannon (1863–1941), who significantly altered the previous scheme into the now familiar sequence O, B, A, F, G, K and M. She developed the definitive Harvard system by rearranging the original groups, omitting some letters, and further subdividing the others by numerical divisions such as A8, A9, F0, F1, . . . She demonstrated that nearly all stars could be categorized into these few spectral types, and that these types could be arranged in a continuous series; she further showed that the spectral behaviour of helium lines could be identified as a continuous sequence of the O, B and A stars, implicitly linking that order in an evolutionary descendency from nebulae.

In 1901, after five years of research, Miss Cannon published a description of the spectra of 1122 of the brighter stars, the foundation work for her larger catalogues. In 1908 the *Revised Harvard Photometry* appeared, containing magnitudes and spectral classes for 9110 stars to magnitude 6.5. Neither Miss Cannon nor Miss Maury was named, but Miss Cannon's spectral classifications were used for all the stars of her 1901 list. Miss Maury's classification was not mentioned, nor was the c-star peculiarity, although footnotes stated that some spectra showed narrow hydrogen lines.

When Hertzsprung received a copy of the work, the omission of the c-star peculiarity caused him concern, for there was now no direct way to identify what he felt was an important stellar luminosity characteristic; and he realized that his earlier letter and papers sent to Pickering had failed to convert the Harvard astronomer to Miss Maury's system. Hertzsprung wrote again to Pickering, trying to convince him of the importance of Miss Maury's identification of c characteristics. In this letter, dated 22 July 1908, Hertzsprung argued that "To neglect the c-properties in classifying stellar spectra I think, is nearly the same as if the zoologist, who has detected the deciding differen-

In the groups K and M the difference between absolute brightness of the stars is perhaps still greater and whatever may be the cause of such difference, it is a priori probable that there will be some marked distinction between the spectra of such stars, f. ex:

	magn.	parall.	diff. in abs. brightness		magn.	parall.	diff. in abs. brightness
α Aurigae	.21	.09	} magn. 4·5	α Tauri	1·06	·12	} 5·9
α Centauri	·06	·76		61 Cygni	4·96	·30	
α Bootis	·24	·03	} 7·6	α Orionis	1·	·03	} 12·1
70 Ophiuchi	4·07	·17		Lal.21258	8·5	·25	

6.2. In a letter of 15 March 1906 to Pickering, E. Hertzsprung paired stars with the c and non-c characteristics, showing the increasing difference in absolute brightness from the K stars (upper left pair) to the M stars (lower right pair).

ces between a whale and a fish, would continue in classifying them together."

Pickering's response in August was firm; he was too sceptical of his own spectroscopic data to believe that characteristics as subtle as those Miss Maury had detected could be real. The objective prism spectra she used could be unreliable because of misalignment, poor focus or variations due to plate flaws that could easily wash out line structure to make it appear anomalous. He felt that such slight variations as Hertzsprung was studying could be found only with high-dispersion slit spectra obtained from the largest telescopes, and not from Harvard's existing equipment. Indeed, Pickering had already said as much in print as early as 1901, when it was originally decided not to continue the Maury classification.

Pickering's scepticism need not surprise us since at the time no one system of spectral classification was universally accepted. Pickering's American colleagues continued to use the systems of Secchi and Vogel in their simplest forms, and there was no guarantee that the vast amount of effort and funds put into the Draper Classification would bear fruit. Any additional system would add to the confusion, and it is possible Pickering was concerned that the adoption of two systems at Harvard would weaken the possibility of the acceptance of either one.

Hertzsprung, of course, could see only that Pickering seemed unwilling to examine a very important element in the work of one of his own staff. Hertzsprung therefore wrote to Pickering again, arguing that it was significant that none of Miss Maury's c stars showed any appreciable proper motions, and further, that Pickering's doubts about the clarity of line spectra could not apply to the observed variation in line ratios within each spectrum; but all to no avail.

In 1909, after Hertzsprung had published another paper reviewing and adding to his work, Schwarzschild invited him to a post at Göttingen, and later asked him to come with him when Schwarzschild assumed the new directorship at the Astrophysical Observatory at Potsdam. Becoming well-established at Potsdam, Hertzsprung continued along similar lines of research, and continued to study the colour and magnitude characteristics of star clusters. In the United States and England, however, his work remained virtually unknown.

At Harvard Pickering pushed ahead with his programme for a more comprehensive catalogue of spectra. In the relatively brief interval from 1911 to 1915, Miss Cannon classified most of the 225 300 stars for *The Henry Draper Catalogue*, but because checking and arranging the material required

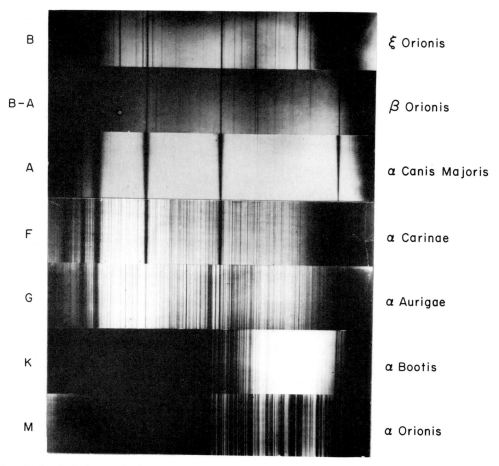

6.3. The frontispiece to Annie J. Cannon's classification of bright southern stars depicting large-dispersion spectra of representative types in the Harvard system.

several additional years, the first volume did not appear until 1918 and the ninth and final only in 1924, after Pickering's death.

To appreciate more fully Pickering's role in the years in which the foundations for the Hertzsprung–Russell diagram were laid, and his general disregard for Hertzsprung's work, we now turn to Pickering's association with Russell.

Prior to 1900, the most important work of Henry Norris Russell (1877–1957) was a short study determining the densities of Algol eclipsing variables carried out while he was still a graduate student at Princeton University. This study, and an unpublished paper on the cooling of a gaseous sphere written around 1901, are indications of Russell's early interests in stellar evolution. He wished to study the evolutionary dynamics of binary star

systems, and so, after his doctorate, he applied for postgraduate study with George Darwin at Cambridge. There, Russell came into contact with the Chief Assistant at the University Observatory, A. R. Hinks (1873–1945), and through this association developed one of the first photographic parallax programmes ever attempted.

But even the pilot parallax programme of Russell and Hinks involved Russell's interests in evolution, albeit by a very indirect route. One would expect that in establishing a new programme utilizing new techniques, Russell and Hinks would choose programme stars that had parallaxes already determined, and would select new candidates on the basis of their capability of showing measurable parallaxes. Proper motion studies already showed that fainter stars were among the best candidates;

6.4. Hertzsprung (upper left) and Russell (centre) in this detail from the group photograph of the International Union for Cooperation in Solar Research, Bonn, 1913.

it is therefore significant that of the fifty-five stars chosen by Russell and Hinks, no fewer than twenty-one also figured in a list recently published by Lockyer (1902) of stellar spectra of the *brightest* stars. Furthermore, most of those stars chosen for study by Russell and Hinks that figured in Lockyer's list fell at the bases of Lockyer's temperature arch, just the places where one would expect stars of similar colour, and hence spectral class, to differ greatly in other characteristics, such as size or luminosity. Russell's lecture notes at Princeton (where he was appointed in 1905), and a letter written to Lockyer in 1911, leave little doubt that Russell was intriguied by Lockyer's theory of stellar evolution, and saw his parallax programme as a possible means of examining its validity.

During the period 1906–9, in order to reduce his parallax measures from relative to absolute values, Russell needed the brightnesses and spectra of both the parallax stars and the comparison stars used in the study. Early in 1908 he came into contact with Pickering, and by the end of April Pickering had set his staff to the determination for Russell of the magnitudes and spectra of some 300 stars. For over a year the Harvard staff worked on this project, and by September 1909 the majority of the stars had been analysed. Russell received the data and on 24 September he wrote back with his first findings:

> . . . *the fainter stars average redder than the brighter ones. I do not know of any previous evidence on this question . . . I would not now risk reversing the proposition and saying that the red stars average intrinsically fainter – some of them certainly do; but Antares and α Orionis are of enormous brightness, and the average may be pretty high.*

Russell's findings were very close to Hertzsprung's. He had identified two sequences in luminosity, and found that the fainter sequence diminished in brightness with advancing spectral class. It seems curious that (if we may judge from the surviving correspondence) Pickering was not prompted by Russell's remarks to mention the similarity of Hertzsprung's conclusions which he had known about for over a year.

Through 1909 Russell continued to work on his parallax results, but he also began to pay more attention to binary stars, their origins, and their brightnesses and masses. By 1910 this work had convinced him that the density range among stars was vast while the mass range was small, and that stars of great dimension existed. With his work on binary stars to support his parallax data, Russell presented his findings at a number of professional meetings, each time calling for a reinterpretation of the normal course of stellar evolution similar to Lockyer's temperature arch. At one of these meetings he met Schwarzschild and learned of Hertzsprung's research; Schwarzschild quickly contacted Hertzsprung, telling him to send his papers to Russell, and by the end of 1910 the two were in correspondence.

The first diagrams

Hertzsprung and Russell came to the diagram from very different directions and with very different interests. Both had established the fundamental relationship – for most stars other than the brightest, absolute brightness decreased with increasing redness – without actually creating an explicit diagram, although Hertzsprung by 1906 and Russell by 1909 certainly had sufficient data to do so. Russell's first graphical exposition appeared in his lecture notes for March 1907, which contained a diagram of stellar temperature decreasing with stellar age. The first actual colour–magnitude diagrams showing apparent magnitude against spectral class were produced by Hertzsprung as early as 1908. They were not published because, it appears, Hertzsprung had detected serious instrumental errors in the colour and magnitude data for the clusters. By 1910 Hertzsprung had solved many of these instrumental problems, and, with the help of an assistant, H. Rosenberg, he published cluster diagrams for the Pleiades and Hyades star clusters.

In the interval between 1909 and 1913, Russell refined his data, primarily through analyses of light curves of eclipsing binary stars carried out in 1912 by Russell's first graduate student, Harlow Shapley (1885–1972), who had come to Princeton specifically to study binary stars under Russell.

Shapley's work, utilizing reduction methods pioneered by Russell, verified that the very brightest yellow and red stars were enormous in radius; and Russell began to realize that the variation in mass from star to star was very small when compared to the observed variation in luminosity, though they varied in the same direction. This relationship between mass and luminosity was later strengthened and clarified through additional observations by many astronomers, including Russell, and in the mid-1920s was discussed theoretically as a general law for stars by A.S. Eddington (1882–1944). This mass–luminosity law will be discussed later in this chapter.

Russell's diagrams first appeared in 1914 in his classic paper "Relations between the spectra and other characteristics of the stars", published both in *Nature* and in *Popular Astronomy*. In this paper Russell acknowledged Ritter's priority in discussing a theory of stellar evolution according to which the stars first heated as perfect gas spheres, then attained densities too great for the perfect gas condition, and then cooled. But one of the immediate problems in interpreting the Main Sequence, or 'dwarf' branch of the 'reversed 7', as the cooling branch of the temperature curve, was that the rudimentary mass–luminosity relationship Russell had detected by 1912 showed that the average masses of stars decreased along the Main Sequence from blue to red. If this was indeed an evolutionary track, as Russell saw it, how did the stars lose mass? Russell felt that the explanation was to be found in Ritter's reasoning that only the most massive stars reached the top of the diagram. Less massive stars would peak at lower temperatures on the giant branch and then join the Main Sequence as dwarfs at a lower point. Thus, only the most massive stars would be at the top, and the average mass would diminish down the Main Sequence, since lower portions of the sequence were statistically enriched with the inclusion of more stars of lower mass. All stars were still envisaged as cooling as they passed down the sequence.

Even though Russell's interpretation of the diagram was not to survive the mid-1920s, the diagram itself remained as an empirical fact. But in the first few years of its existence, many astronomers still were sceptical of the vast range in stellar dimensions it revealed. Two important

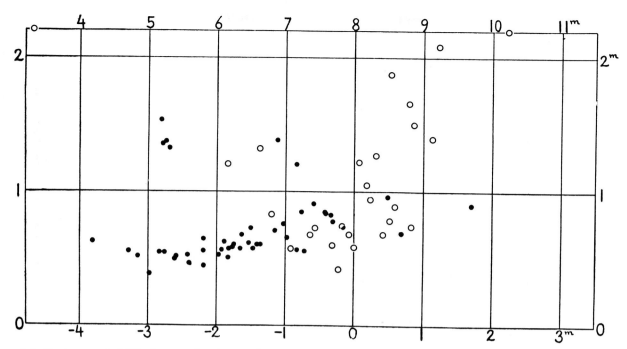

6.5. Hertzsprung's 1911 diagram for the Hyades star cluster. The horizontal coordinate is apparent photographic magnitude, the vertical coordinate is a measure of stellar colour. Open circles are probable members of the cluster.

events helped to change this. The first was the development and use of spectroscopic parallaxes, and the second was the successful measurement of the angular diameter of a star in 1920.

Spectroscopic parallaxes

The use of line ratios and line intensities in stellar spectra to estimate the intrinsic luminosities of stars had its origin in Lockyer's attempts to find spectroscopic criteria that would help him assign a star to the rising or falling temperature branches. A successful procedure developed out of Miss Maury's work; Hertzsprung extended it by showing that the c and non-c subclassifications in fact related to luminosity, but he did not carry the procedure through. By 1910, both Hertzsprung and Russell were calling for spectroscopic studies that would provide luminosity information, and Hertzsprung, during a stay as a guest investigator at Mt Wilson in 1912, encouraged W.S. Adams to look for luminosity effects.

Kapteyn, who had been instrumental in bringing Hertzsprung to Mt Wilson, also provided unwitting aid by suggesting to Adams and A.

Kohlschütter that they place the spectra of high and low proper motion stars of the same spectral class on the same photographic plate to minimize photographic errors in their search for effects of interstellar absorption, such a search being a longstanding passion with Kapteyn. By early 1914 Hale reported to Kapteyn that the method seemed to show the spectroscopic effects of interstellar absorption. At first, Adams agreed that the line and continuum brightness differences seen in the spectra of the two stars on each plate were evidence for both selective and general absorption. Adams was aware, however, that the spectral differences could be due to luminosity differences, and Kapteyn wondered why the line absorption differences were also a systematic function of spectral type. Still, both Hale and Kapteyn preferred the interstellar absorption interpretation. But in May Hertzsprung wrote to Adams reminding him of the possibility of luminosity effects, and before the end of the month Hale wrote to Kapteyn:

Has Adams written to you regarding his remarkable new result, the most important of all? He can now

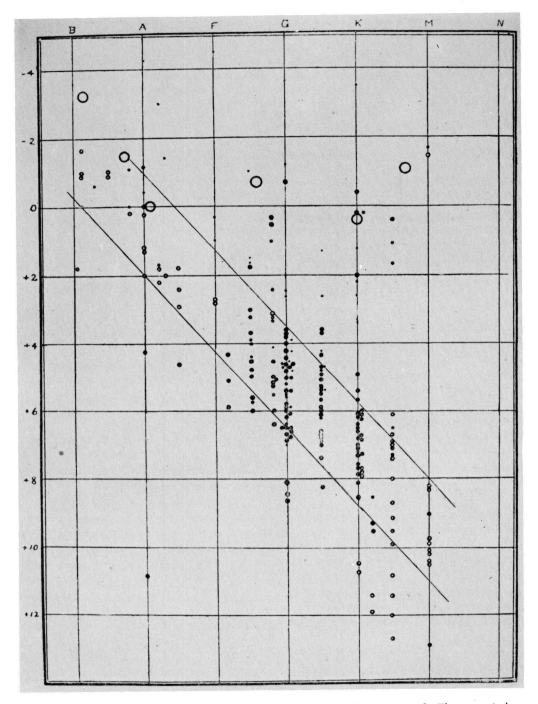

6.6. The first published 'Russell diagram' correlating spectral type and absolute magnitude. The open circles are mean absolute magnitudes for bright stars with small parallaxes. This version appeared in *Nature* in 1914, and simultaneously in *Popular Astronomy*.

calculate the parallax of a star with apparently high precision from the apparent magnitude and the ratio of intensities of certain lines of the spectrum!

Hale rejoiced in Adams's powerful new technique of luminosity and distance determination. This technique of 'spectroscopic parallaxes' provided a vastly larger sample of evidence for the existence of giants and dwarfs. In May 1916 Eddington wrote to Adams congratulating him on a job well done and said, "I think the question of the giant and dwarf division of red stars must now be considered finally settled." Nevertheless, Eddington had written in March to Lockyer admitting some continuing difficulties in Lockyer's (and now Russell's) theory of ascending and descending temperature stars; but he had concluded by saying that "I believe that in the main it must be right. I find myself even in the last year advancing towards that point of view."

The angular diameter of Betelgeuse

Confident as he was concerning the existence of giant stars, Eddington still realized that direct observation of the diameter of a giant star was essential to prove that the indirect spectroscopic method and the predictions of black body radiation laws were correct. In the autumn of 1920 he provided several explicit calculations, based upon theory, for what the observed angular diameters of some giant stars might be. Eddington was fully aware that an instrument capable of making the desired measurement was nearing completion at Mt Wilson.

For nearly thirty years, the physicist A. A. Michelson (1852–1931) had been interested in using interferometric methods to determine the angular diameters of stars. Successful tests in the early 1890s on the Jovian satellites foreshadowed eventual success for stars. But the project was stalled for several reasons, notably because at the time there was no way to predict reliably the angular diameter of a star, and therefore Michelson could not determine what size interferometer was needed to make the measurement. As his 100-inch (254-cm) reflector was nearing completion after World War I, Hale decided that a Michelson interferometer would be an excellent auxiliary instrument for the giant telescope, useful for studies of close double stars even if it could not measure the angular diameters of stars.

When Eddington's predictions appeared, the interferometer was still being tested and was not yet fully operational. The Mt Wilson staff hastily prepared the instrument, taking several short-cuts in the design. One short-cut required an observing assistant to sit on top of the interferometric cross-bar to move the interferometric mirrors bit by bit during the tedious observations. Finally, after hurried tests, and several difficult nights of fine-tuning the instrument, F. G. Pease and J. A. Anderson succeeded in measuring the angular diameter of the star Betelgeuse on 13 December 1920. Its angular diameter was 0".047, just slightly less than Eddington's prediction.

The existence of stars of vast dimension had been confirmed. For Betelgeuse to subtend an angle of 0".047 and have a parallax thought at that time to be 0".018 meant that it had to be larger than the orbit of the Earth!

The fate of Russell's theory 1914–26

As he was writing his words of congratulations to Adams, and the existence of giants and dwarfs was becoming generally acknowledged, Eddington himself was pursuing the application of radiative equilibrium to the theory of the interiors of stars.

Virtually all late-nineteenth-century theories of stellar structure assumed that heat was transported from the interior to the surface of a star by convection currents. Lane had been much influenced by the lectures of J. P. Espy on convection in the Earth's lower atmosphere, and Ritter and Emden had clearly related their work to terrestrial meteorological phenomena. The first to suggest that heat was transported by radiation through the solar atmosphere was R. A. Sampson in 1894. In the period 1902–5 Arthur Schuster examined the role of pure absorption and scattering in the solar atmosphere; and in 1906, aided by recent advances in thermodynamics, Schwarzschild developed more fully the idea that energy transport through an atmosphere was predominantly by radiation rather than by convection, and was successful in creating an atmosphere that reproduced the observed limb-darkening of the Sun. But the old idea of convective transport was hard to overcome, partly because the visible surface of the Sun (the photosphere) resembled the surface of a boiling liquid complete with convective cells.

Eddington's revival of Schwarzschild's work

extended the theory of radiative equilibrium to the stellar interior. In 1916 Eddington applied Schwarzschild's formulae, adapting them to account for the curvature of successive layers in the interior of a star, where the radius of curvature became an important consideration. At first, Eddington's efforts were wholly in the direction of determining the maximum temperatures attainable by stars of different masses. Clearly guided by Russell's temperature arch theory, Eddington felt that radiative equilibrium applied to gaseous giant stars but not to dwarfs. The steep temperature gradient derived from his calculations of the structure of giant stars that resulted in the dominance of radiative equilibrium simply was not thought to be found in dense dwarfs. So, while Eddington considered that dwarf stars deviated from the perfect gas condition, he did note that what he considered to be extreme deviations from the perfect gas state seemed to produce very small corrections to the assignment of the maximum temperature attainable by a 1.5 solar mass star.

As Eddington pursued the problem, however, other advances were being made. The most significant were the development of the Bohr theory of the atom culminating in (1) M. N. Saha's studies around 1920 of ionization equilibrium in the atmosphere of stars, (2) the recognition by Jeans, J. Eggert and others that a high degree of ionization exists in stellar interiors greatly reducing their mean molecular weight, and (3) the work of H. A. Kramers on stellar opacities. By 1924 Eddington had realized that the condition of high ionization in the stellar interior allowed all stars (except white dwarfs) to maintain the perfect gas state throughout their lives.

Eddington developed a theoretical discussion of the mass–luminosity relationship in a paper in 1924 in *Monthly Notices of the Royal Astronomical Society*, the culmination of work undertaken by Russell, Hertzsprung and others since 1911. Eddington first thought that the relationship would hold only for stars in the perfect gas state and therefore only for giants. But in trying to explain why stars of similar mass could exist with widely differing luminosities, he found to his surprise that both giants and dwarfs satisfied the perfect gas equations. He had expected to see the dwarfs fall well below the mass–luminosity curve because of their assumed incompressibility; since they did not,

they too had to behave as perfect gas spheres, and hence be perfectly compressible. These two discoveries meant that Russell's theory of evolution down the Main Sequence was not tenable. If the stars there were still in the perfect gas state, then continued contraction would cause them to heat up and not to cool.

Eddington was concerned that his results were in "conflict with the Lane–Ritter theory of stellar evolution as incorporated in the giant and dwarf theory at present almost universally accepted". He added that "strong initial opposition to the results in this paper will doubtless be felt on that account". To deal with the expected opposition, he included in his paper a discussion of the observational aspects of the giant and dwarf theory of evolution. As Eddington saw it, the critical opposition would come from the observed fact that the average masses of stars became less as one moved down the Main Sequence. No star could pass along the 'reversed 7' path from being a red giant to becoming a red dwarf because the masses were different for those two classes of stars. Eddington's work had shown that the mass difference alone could account for the observed luminosity difference between those classes, and therefore the only way to link them in an evolutionary sequence was to find some way to reduce the mass of a star as it aged. If, somehow, the source of stellar energy included mass annihilation in addition to gravitational contraction then mass loss could occur. But one could not sustain the giant and dwarf theory on the basis of speculation concerning energy sources:

The difficulty is merely that we have no knowledge of the conditions under which the star's unknown source of energy is liberated; and in these circumstances speculation is unlimited and chiefly unprofitable.

Eddington's anticipation of opposition resulted from his on-going debates with James Jeans (see Chapter 16). In this particular instance, however, it was conceivable that Russell might become defensive since his theories were now being questioned. But in 1923 Russell had created much of the data, and had helped to produce (with Adams and A. H. Joy) observational material to which Eddington could point as agreeing with his own conclusions. Russell had also been following theoretical developments in stellar structure, and during these years had contributed commentary to the literature on the role of radiative equilibrium. He was, therefore, openly receptive to Eddington's findings, and sympathetically reviewed the new situation in 1925 in a paper in *Nature* entitled "The problem of stellar evolution", concluding that: "Several investigators – Jeans, Kramers, Eggert – have contributed to this field, but much the largest share is Eddington's."

By 1927, Russell had abandoned his interpretation of the Main Sequence as an evolutionary track, and he now looked for other possibilities. To Russell, as well as to most other workers in this area, the largest single problem that remained was the determination of the source of stellar energy and the resolution of the question of why giants exist at all.

The fate of Russell's theory was sealed by the time Eddington's famous book, *The Internal Constitution of the Stars*, appeared in 1926. Eddington assessed Hertzsprung's and Russell's more or less independent routes in coming to the diagram, and then remarked on the status of the evolutionary ideas:

In speaking of the 'overthrow' of the giant and dwarf theory, we do not intend to imply that there is to be any retrogression with regard to these results. They are mostly the ascertained facts, some of which the authors discovered at the time that they theorised on them; others followed by natural development. In any reconstruction of theory facts must be attended to. But the facts were welded by an attractive theory of evolution on the lines of Lane's and Lockyer's earlier proposals into a remarkably coherent scheme. I do not think it is too blunt an expression to say that this is now overthrown; at least it has been gutted, and it remains to be seen whether the empty shell is still standing.

Epilogue

Throughout this chapter, we have examined theories of the evolution of the Sun and stars within the context of gravitational contraction as the sole source of stellar energy. We have noted only in passing that strong indications eventually appeared of the necessity to include non-gravitational sources. Consideration of gravity as the source of stellar energy allowed rational schemes to be developed for stellar structure and

evolution, but it fell far short of providing the solar duration demanded by geology. At the turn of the century Thomson had almost succeeded in his life-long struggle to convince everyone that the original geological time scales were excessive. It was only with the discovery of radioactivity that his arguments for short time scales collapsed. During the interval considered in this chapter, astronomers, unsure of the physical nature of the Sun and stars, remained on the sidelines of the time-scale debate; they did not pursue the time-scale problem as a central issue, and when they did discuss the matter, more often than not Thomson's arguments were repeated. The larger effort in astronomy was towards the acquisition of good useful data, and with this data came the irresistible urge to speculate. While the theories of Ritter, Lockyer and Russell were certainly premature, Russell's gained wide acceptance because of his better data and the acknowledged existence of giant stars.

Ironically, recognition of the very existence of giant stars revived the stellar energy question. When Eddington first examined the problem of stellar structure in 1916, he was interested in the question of stellar stability rather than in the ages of stars. But as his research progressed, and he began to incorporate knowledge of the mass–luminosity law into his work, the problem of stellar energy became acute. Giant stars radiated energy so profusely that to limit the source of energy to gravity alone restricted their ages to less than 100 000 years. Eddington thus began to include the idea that some form of radioactive source of energy supplemented gravitational energy, though he left its exact nature open. But Eddington's arguments for the role of radiative equilibrium and the possibility of non-gravitational energy sources did not go unopposed. Prominent in opposition was Jeans, although many others joined the discussion from time to time. The controversy between Eddington and Jeans, and the search for the source of stellar energy, will be covered in Chapter 16.

Further reading

Agnes M. Clerke, *Problems in Astrophysics* (London, 1903)
David H. DeVorkin, A. A. Michelson and the problem of stellar diameters, *Journal for the History of Astronomy*, vol. 6 (1975), 1–18

David H. DeVorkin, The origins of the Hertzsprung–Russell diagram, in A. G. Davis Philip and David H. DeVorkin (eds), *In Memory of Henry Norris Russell* (Dudley Observatory Report no. 13, 1977), 61–77

David H. DeVorkin, Community and spectral classification in astrophysics: The acceptance of E. C. Pickering's system in 1910, *Isis*, vol. 72 (1981), 29–49

Ejnar Hertzsprung, Zur Strahlung der Sterne, ed. by D. B. Herrmann, *Ostwalds Klassiker der exakten Wissenschaften*, no. 255 (Leipzig, 1976).

D. Hoffleit, The discovery and exploitation of spectroscopic parallaxes, *Popular Astronomy*, vol. 58 (1950), 428–38, 483–501; vol. 59 (1951), 4–19

Karl Hufbauer, Astronomers take up the stellar-energy prob-lem, 1917–1920, *Historical Studies in the Physical Sciences*, vol. 11 (1981), 277–303

A. J. Meadows, *Science and Controversy: A Biography of Sir Norman Lockyer* (London, 1972)

Axel V. Nielson, Contributions to the history of the Hertzsprung–Russell diagram, *Centaurus*, vol. 9 (1963), 219–53

J. Scheiner, *Die Spectralanalyse der Gestirne* (Leipzig, 1890), transl. by Edwin Brant Frost as *A Treatise on Astronomical Spectroscopy* (Boston, 1894)

Bancroft W. Sitterly, Changing interpretations of the Hertzsprung–Russell diagram, 1910–1940: A historical note, *Vistas in Astronomy*, vol. 12 (1970), 357–66

PART II
Observatories and instrumentation

7

Astronomical institutions

Introduction

Owen Gingerich

In 1886 A. Lancaster tabulated over two hundred observatories throughout the world, distributed as follows:

Europe	150	Asia	5
North America	42	Africa	5
Latin America	15	Australia	4

In this chapter we can mention only a few of these observatories; there are, however, several institutions which, by the size of their telescopes or by the vision of their directors and staff, played such an outstanding role in the development of astronomy that no serious account of the history of this science could afford to neglect them. In the sections that follow, seven of the most important of these will be examined in detail from the point of view of their instruments and their administrative arrangements. While their scientific goals are of course mentioned, their scientific achievements are in general left to other chapters. In addition, Chapter 8 describes the observatories organized by George Ellery Hale, while Chapter 9 takes up a specific discussion of the southern-hemisphere institutions.

Of the four major European observatories considered in some detail here (Greenwich, Paris, Pulkovo, Potsdam), all are national institutions supported by their respective governments; in contrast, of the six American observatories discussed in the subsequent sections of this chapter and the following (Harvard, US Naval, Lick, Yerkes, Mt Wilson, Palomar), one (and only one) originated from national funding, another as a university observatory, and the others primarily through private philanthropy. The government-supported institutions generally had specifically defined missions (such as time service and epheme-

rides), although, as these accounts show, powerful directors could guide the research into new areas; the privately supported observatories, functioning as major research centres, depended even more on the guidance of their directors (and their skill as fund raisers!) for the development of their programmes. In general, these private observatories were relatively wealthy and could command the greatest instrumental resources. In 1920 these ten observatories owned four of the six refractors of 30-inch aperture and above (30-inch refractors being also at Nice and Allegheny), and four of the six operating reflectors of 40-inch aperture and above (a 44-inch reflector being at Flagstaff and a newly-completed 72-inch reflector at Victoria).

There existed, however, a vast network of smaller observatories that contributed significantly to specific problems despite their small number of personnel. The Zurich Observatory, established in 1864 through the efforts of Rudolf Wolf, long continued his standard work on sunspot numbers and statistics. The observatory of the university in Bonn, founded almost singlehandedly by F. W. A. Argelander, became famous for the *Bonner Durchmusterung* (a catalogue of nearly one-third of a million stars) despite the modest size of its staff.

Many of the smaller observatories were originally the creation of wealthy or enterprising amateur astronomers, and in many instances these were eventually taken over by educational institutions. The situation at the Cambridge Observatories is an excellent case in point. Founded in 1820 by the Senate of Cambridge University, it was at first funded half by the University and half by public subscription; the University provided a director through one of its professorships, and two assistants, but the acquisition of new instruments came predominantly from private endowments or by direct gift. In this way the institution acquired in 1890 the principal telescope from R.S. Newall's private observatory, a 25-inch (64-cm) refractor; in 1904 the 12-inch (30-cm) photographic refractor from F. McClean's private observatory came as well; and in 1908–09 the telescopes formerly used in William Huggins's observatory were transferred to Cambridge. Furthermore, in 1911 the Solar Physics Observatory, which had been founded by Norman Lockyer and which was then in South Kensington (London), was moved in its entirety to Cambridge to a site adjacent to the existing observatory. For 35 years this observatory continued a separate existence with its own director, staff and library, and not until 1946 was an amalgamation of the two institutions accomplished.

The mixture of public and private observatories is well illustrated by the situation in Vienna in the latter part of the nineteenth century. The older royal observatory, established at the university in 1756, made rather few observations of value owing to its unsuitable site, and remained comparatively undeveloped. Thus it was easily surpassed by a scholar of private means, T.v. Oppolzer, who established in 1865 in the outskirts of Vienna a private observatory with a refractor larger than any other in the Austrian empire, with a 7-inch (18-cm) aperture. Soon thereafter, Professor K.L.v. Littrow drew up plans for a major university observatory at a new site in the Vienna suburbs. By 1872 the plans were sufficiently advanced that his protégé E. Weiss was sent on a tour of observatories in Britain and the United States. As a consequence, a 27-inch (69-cm) refractor was commissioned from Grubb in Dublin; mounted in 1881, by which time Weiss had become director of the observatory, the refractor was for a few years the largest in the world. Oppolzer's evident abilities in celestial mechanics and geodesy had in the meantime won him a full professorship at the university (in 1875), so he headed an entirely separate institute of theoretical astronomy, but which was without its own observatory. This dual arrangement continued at the University of Vienna until very recently.

In 1884, soon after the large refractor was installed at the Vienna Observatory, M.v. Kuffner (1854–1939), a wealthy Jewish brewer who had been inspired by Oppolzer, decided to found his own observatory in a suburb of Vienna with one of Oppolzer's assistants as director. To it he brought the world's largest heliometer (by Repsold, 21.7 cm, 1894) as well as the largest meridian circle in the Austrian empire (by Repsold, 8.1 cm, 1893). The Kuffner Observatory carried out a significant part of the southern extension of the Astronomische Gesellschaft zone catalogue, from −6° to −10°. The excellence of such a private observatory can be seen in the quality of its staff, which included, on short appointments, K. Schwarzschild, J.F. Hartmann and G. Eberhard.

In contrast with the situation in Europe, few of the major American observatories were privately owned, although many college or university observatories were supported by individual benefactors. For example, the Washburn Observatory at the University of Wisconsin was founded in 1878 and its 15½-inch (39-cm) refractor was installed through the generosity of Governor C.C. Washburn; at the University of Virginia the observatory with its 26-inch (66-cm) refractor was made possible through the benefaction of the industrialist Leander McCormick in 1882. Public subscriptions led to the formation of other observatories, such as those founded at Cincinnati, Ohio, in 1842, at the University of Michigan around 1853, and in Pittsburgh (the Allegheny Observatory) in 1860.

This means of finance occasionally led to some differences in the perceived roles for such institutions. Conflict between popular and professional goals is nowhere better illustrated than in Albany, New York, in the early years of the Dudley Observatory. This owed its origin to the enthusiasm of an itinerant astronomical lecturer, Ormsby McKnight Mitchel, who in 1851 inspired wealthy citizens to found their own observatory. Meanwhile, a group of American savants, the self-styled 'Scientific Lazzaroni', had been endeavouring to establish in Albany a national graduate university with European standards. They seized upon this opportunity to incorporate the observatory into their plans; after the observatory had been erected in 1854 and it had become clear that Mitchel intended to remain as director of the similarly-founded popular observatory in Cincinnati and was not willing to come to Albany, the Lazzaroni offered to have one of their own members, B.A. Gould, assume charge of the observatory – at no cost to Albany, since he and his corps of assistants were already paid by the US Coast Survey.

Gould had studied in Germany with Gauss and was the founder of *The Astronomical Journal*. Having little patience with popular astronomy and planetary observing, he attempted to impose his own high professional standards on the new Dudley Observatory. For example, he demanded that the piers for the 8-inch (20-cm) meridian circle be the largest in the world, at the expense of tearing down a wall already built. However, the delays in obtaining the instruments, Gould's frequent absence from Albany, and his firm opposition to admitting the public to the observatory grounds exasperated the trustees; a fierce pamphlet war broke out, and eventually (in 1859) a band of local toughs, probably hired by one of the trustees, broke into the observatory and forcibly evicted Gould. After this regrettable episode Gould continued his distinguished astronomical career elsewhere, founding the Argentine National Observatory (see Chapter 9); in the present century the Dudley Observatory made its mark as the home of the *General Catalogue* of star positions (see Chapter 13).

In the sections that follow, the scope of the discussion is generally limited to the period 1850 to 1920. Nevertheless, it may be appropriate here to mention the observatory that for some years had the world's second largest telescope, even though its foundation occurred after 1920. In common with many of the American observatories, the McDonald Observatory was established through private munificence, from a bequest by W.J. McDonald to the University of Texas in 1926. Administrative arrangements between the Universities of Texas and of Chicago for operating the observatory were established in 1932, and in 1939 an 82-inch (208-cm) reflector was dedicated on Mt Locke in western Texas.

Greenwich Observatory
Philip S. Laurie

By the middle of the nineteenth century the Royal Observatory at Greenwich was one of the world's leading astronomical institutions with a reputation in fundamental astronomy second to none. Under the direction, since 1835, of the seventh Astronomer Royal, G.B. Airy (1801–91), the Observatory had been enlarged and provided with

instruments of his own design, and had entered into a strict regime totally dominated by the Astronomer Royal. The installation of the Transit Circle in 1851 and the introduction of time dissemination by the electric telegraph used by private companies and railways (which led to the adoption of Greenwich Time for the railway system in 1852) were closely followed by the provision of a $12\frac{3}{4}$-inch (32.4 cm) refractor. Airy was able to state in his annual report for 1859 that

the inauguration of the New Equatorial will terminate the entire change from the old state of the Observatory. There is now not a single person employed or instrument used which was there in Mr Pond's time, nor a single room used in the Observatory which is used as it was then.

The Royal Observatory had been under Admiralty control since 1818, the Hydrographer of the Navy having a special responsibility and being, in effect, Airy's link with higher authority. A Board of Visitors, comprising the President and five Fellows of the Royal Society, a similar number from the Royal Astronomical Society, the professors of astronomy at Oxford and Cambridge, and the Hydrographer, nominally supervised but actually worked with the Astronomer Royal to promote the efficiency of the Observatory in domestic and international fields. The Admiralty bore the running expenses, which in 1860 came to £4300, including Airy's salary of £1000. The permanent staff numbered only seven assistants, the First Assistant being Robert Main (1808–78), a mathematician of repute who acted as Airy's confidential aide. His salary was £500 a year; remaining staff salaries and allowances totalled £1200, ordinary expenses (taxes and repairs) amounted to £1000, while 'extraordinary' expenses (instrument modification and purchase) came to a further £600. Supernumerary computers were employed on a 'hire-and-fire' basis when large-scale reductions were required, their combined salaries costing a few hundred pounds annually.

At the annual Visitation (usually held on the first Saturday in June) the Astronomer Royal read a report to the Board of Visitors, enumerating the activities of the past year in detail and making recommendations for improvements. When agreed, the latter were submitted to the Admiralty for the necessary funding. These annual reports were printed with the astronomical, magnetic and meteorological results in the Observatory's principal publication, the *Greenwich Observations*.

In addition to these regular activities, Airy undertook many long-term projects, such as the *Reduction of Observations of Planets, 1750–1830* (published 1845) and *Lunar Reductions 1750–1830* (1848), although much of his time was also taken up with extraneous Governmental business. His last two decades saw further expansion at Greenwich, including the beginning of astrophysical studies and, from 1873, long-term photographic observations of the Sun, the photo-heliographic and spectroscopic results being incorporated in the annual volume. Thus the Observatory was a pioneer in observational astrophysics and solar physics. Also it took a lead in the development of the geophysical sciences, starting with geomagnetic observations in the 1830s, meteorological observations soon thereafter, and in 1854 going on to a measurement of the mean density of the Earth.

Airy was the last Astronomer Royal to be appointed by Royal Warrant (all his successors being civil servants), and during his long term of office the Royal Observatory evolved from a famous but specialized observatory into a great national institution from which those in authority sought advice. He was succeeded in 1881 by W.H.M. Christie (1845–1922), the first staff member to attain the office of Astronomer Royal. In 1880, Greenwich Mean Time had been adopted as the legal time for Great Britain, and four years later Greenwich was adopted as Prime Meridian at the Washington Conference.

The need for greater light-grasp led Christie to press for a much larger telescope, and a 28-inch (71-cm) refractor was installed in 1893. Meanwhile, funds for a 26-inch (66-cm) photographic refractor were presented in 1894 by Sir Henry Thompson, a surgeon, who also provided for a 30-inch (76-cm) reflector. The cruciform building, known at first as the New Physical Observatory and the largest structure at Greenwich, was built between 1891 and 1899; it provided much-needed office and laboratory accommodation in addition to housing the 26-inch and 30-inch telescopes on a twin mounting. Christie also obtained a new site for magnetic and meteorological equipment in Greenwich Park, a few hundred metres to the east, but he built his new altazimuth telescope on the

main site in 1898.

In 1890 a 13-inch (33-cm) astrographic refractor was added to the Observatory's equipment in order to participate in the international *Carte du Ciel* plan for a photographic survey of the whole sky down to stars of magnitude 14. This was just one of the many ways in which the Observatory kept in the forefront of international cooperation in astronomy. Both the Astronomers Royal after Christie, Dyson and Spencer Jones, served as President of the International Astronomical Union.

By 1900 Christie and the Board of Visitors had been able to convince the Admiralty of the need for a second Chief Assistant; the rest of the staff comprised five Assistants, two Higher Grade Established Computers, six Established Computers, plus some dozen supernumeraries. The salary bill for the Astronomer Royal and the scientific and ancillary staff came to £7700 while contingencies amounted to £1300. The Astronomer Royal was fortunate in recruiting distinguished Chief Assistants. His first was E. Dunkin (1821–98), who was promoted from the staff and retired in 1884. The others, with their terms of office, were: H.H. Turner (1884–94), later Savilian Professor of Astronomy at Oxford; F.W. Dyson (1894–1906), who became Astronomer Royal for Scotland and, later, Astronomer Royal; P.H. Cowell (1896–1900), Superintendent of the Nautical Almanac Office; and A.S. Eddington (1906–13), who became Plumian Professor of Astronomy at Cambridge.

Christie put his equipment to much important use for photography and he led several solar eclipse expeditions to obtain large-scale photographs of prominences and the corona. He encouraged special researches, such as the investigation of the motion of Halley's Comet, carried out by Cowell and A.C.D. Crommelin.

One great misfortune befell towards the end of Christie's time, namely, the building in 1906 of a large generating station (for the London tramways) on the bank of the River Thames only one kilometre due north of the Transit Circle. A protest by the Board of Visitors did something to modify the installation of large reciprocating engines, with their heavy vibrations, but the heated air and pollution from the tall chimneys were to plague the observatory until its removal more than forty years later to the present site at Herstmonceux in Sussex.

7.1. Royal Observatory, Greenwich. The New Physical Observatory in course of construction, March 1897, before the erection of the West Wing. The dome, originally made for the 24-inch Lassell reflector (whose discarded mounting appears at the right), was transferred here to house the 30-inch reflector.

The first Astronomer Royal to retire at the age of sixty-five under the Civil Service rules, Christie was succeeded in 1910 by his former Chief Assistant, F. W. Dyson (1868–1939). The Observatory then had an established staff of twenty, plus twenty-five supernumeraries, led by Eddington and, from December, S. Chapman (1888–1970). The running costs had risen slowly to about £10 500, a figure that remained fairly constant until the early 1920s.

Dyson's plans for new programmes of investigation on the regions covered by the astrographic charts, including photographic magnitudes, colour temperatures, proper motions and parallaxes, had mostly to wait until after the disruption caused by World War I and the return of staff from active service or secondment. During the war, Dyson kept all essential work going at Greenwich, in spite of additional commitments to chronometer

and binocular testing. With the return of staff to pre-war complement, he was able by 1920 to turn his attention to improvement in the technique of time-keeping and dissemination by the introduction of free-pendulum clocks and the use of radio for the benefit of world shipping. He played a considerable part in the formation of the International Astronomical Union and was one of its first vice-presidents.

An indefatigable organizer of solar eclipse expeditions, Dyson was quick to realize the implications of Einstein's theory of general relativity and the feasibility of a practical demonstration at the total eclipse of 29 May 1919. This resulted in the joint Greenwich–Cambridge expedition to Sobral in Brazil and Principe off the southwest coast of Africa, led by Crommelin and Eddington, respectively; Dyson and Eddington afterwards demonstrated that the results were in good agreement with Einstein's prediction. They announced their momentous findings at a joint meeting on 6 November 1919 of the Royal Society and the Royal Astronomical Society presided over by Sir J.J. Thomson.

In the period under review the astronomers of the Royal Observatory harnessed the unprecedented technological developments of the times to the service both of astronomy itself and of the community as a whole. The products of the industrial revolution led to telescopes that were better engineered than any earlier instruments; electrical science and its later application in radio engineering led to the distribution of accurate time by essentially the same means that are still employed; photography was made to serve astronomy in ways that enormously extended the capabilities of its other equipment. Airy possessed an astonishing instinct in such matters: to name only two further examples, he showed how a magnetic compass could be used dependably in an iron ship, and he gave detailed technical assistance with the construction of the Westminster Clock. Christie and Dyson possessed perhaps more sense of direction in the progress of astronomical science for its own sake, but they too were notable pioneers in exploiting new techniques.

Under these three directors, the Royal Observatory became the prototype national establishment for scientific research and development and Airy himself was the prototype government scientific adviser. If it be asked how an institution with such a small staff and such modest equipment could be so productive and influential, it may be remarked that in its Board of Visitors and in the Admiralty it had access when needed to the best available scientific and administrative support. At the same time, it was the very model of an institution run with the absolute minimum of interference by the authorities that were there to support it.

Paris Observatory
Jacques Lévy

When the Bureau of Longitudes was founded in 1795, the administration of the Paris Observatory was among its responsibilities. Four of its ten members had the title of Astronomer, and the Bureau appointed four Adjoint Astronomers.

These eight were formally in charge of the Observatory's instruments, but in practice each worked on whatever took his fancy, so that on occasion the member in charge of administration found himself the only one involved in observations. Even the appointment of D.F.J. Arago (1786–1853) in 1834 as Director of Observations made little difference. It was only with U.J.F. Le Verrier (1811–77) (the first since Cassini IV to receive the title of Director of the Observatory) that there was a reform of the work pattern; the previous freedom, which had bordered on licence, was now replaced by a rigidity of control, and conflicts between the director and the staff resulted in the replacement of Le Verrier by Delaunay in 1870. On Delaunay's death two years later, Le Verrier was reinstated, but the responsibilities of each member of staff were now formally laid down. The staff consisted of six Astronomers and ten Adjoint Astronomers, together with assistant astronomers. (In 1927, in Paris and Meudon, there were seven assistant astronomers and nine calculators and assistants.)

To the main building of the Observatory, which had been little altered since its foundation, two wings were added: an east wing in 1832 for meridian instruments, and a west one for an auditorium where, from 1846, Arago delivered his celebrated public lectures. The main entrance, which had been to the east, was now on the north, between two pavilions built by the architect Vaudoyer when Observatory Avenue was driven through in 1811. With the construction of the two domes over the upper floor in the middle of the century, and the enlargement of the grounds in 1882 by the acquisition of 9000 square metres, the Observatory took on an appearance close to that of the present day.

The men appointed by the Bureau of Longitudes to manage the Observatory were, in turn, J.-J.L. Lalande in 1795, P. Méchain in 1801, J.B. Delambre in 1804, A. Bouvard in 1822, and finally Arago from 1843 to 1853. The Observatory, which had been poorly furnished with instruments, was suitably re-equipped: a complete mural circle of aperture 11 cm fitted with six microscopes was installed in 1822, a 15-cm transit instrument in 1834 and a 12-cm meridian circle in 1843, these last two instruments being by Gambey. The refractor of 38-cm aperture and 9-m focal length intended for the east dome was ordered by Arago but came into service only after his death in 1853.

In the absence of proper leadership, the observations lacked regularity and were not even reduced. On the other hand some important researches were carried out in physics: in the period beginning in 1811 Arago discovered chromatic polarization and rotatory polarization, and with J.-B. Biot he established the laws of rotatory polarization in 1812 and 1815. J.B.L. Foucault likewise worked in the Observatory, and it was within a single room there, in 1850, that he established the velocity of light by the method of mirror rotation, at the same time demonstrating that the velocity of light in water is greater than the velocity in air and so confirming the wave theory of light. In 1856 Foucault also completed the first silver-on-glass mirrors.

It was Arago who first directed the interests of Le Verrier into celestial mechanics, and he later defended Le Verrier's claims to the discovery of Neptune in 1846; but relations between the two then deteriorated and after 1847 Le Verrier made his plans for astronomical research known directly to the government, thereby in effect making a bid for the directorship of the Observatory. The post of Director was indeed established for him after the death of Arago, the annual budget being doubled to the value of 97 500 francs.

Now aged 43, Le Verrier was already professor in the Faculty of Sciences, member of the Institut, a senator, and a scientist of world renown and an unchallenged authority within the Observatory. Yet what he had achieved – the complete theory of the planets – was through his personal efforts alone. Authoritarian and exacting, to him his collaborators were mere underlings. Under his 'rule', the staff did not engage in research. Instead they observed and they calculated, reducing the meridian observations accumulated since 1800 and making meridian observations of the 50 000 stars in Lalande's catalogue.

By 1870 Le Verrier could congratulate himself on having been responsible for the making of more than 200 000 observations and their subsequent reduction. But the staff, who as individuals were forever being harassed and substituted, resigned as a body, and this led to the removal of Le Verrier. His successor was C.-E. Delaunay (1816–72), whose commitment to development of the theory of the

Moon had been the object of violent and unjustifiable attacks on the part of Le Verrier. But Delaunay died at sea in 1872, and because he too had become involved in conflict with his staff and in view of international criticism of the replacement of Le Verrier, the latter was reinstated at the head of the Observatory. Now a sick man, and the wiser for his experience, Le Verrier set himself quietly to the task of completing the tables of Saturn and the theory of Neptune. He died on 23 September 1877, the anniversary day of the discovery of Neptune.

Le Verrier had also been responsible for setting up the first international meteorological service. With the agreement of the telegraphic authorities, in 1858 he began to publish a daily bulletin giving the pressure, temperature and wind direction for some twenty stations. Then, in 1863, he inaugurated a system of storm warnings by means of telegraphic announcements along the coast, at first limited to France but afterwards extended to the ports and capitals of Europe.

It was only under the directorship of Admiral E. B. Mouchez, from 1878 to 1892, that the work of the Observatory began to broaden out. A photographic laboratory was set up as soon as he arrived. The researches of the brothers Paul and Prosper Henry (see Chapter 2) led to the construction of a double photographic equatorial. This instrument was adopted for the *Carte du Ciel* project, which resulted from the international congress summoned by Mouchez in 1887, the eighteen participating observatories agreeing to install similar instruments.

Mouchez also ordered the construction of two coudé equatorials after a design by M. Loewy; here the objective is placed at one end of the declination axis about which it can be turned, and the image, after reflection by two plane mirrors, is formed at one extremity of the polar axis. The first coudé, 27 cm in aperture, was installed in 1883. The second, of 18-m focal length, was installed by 1890; it had interchangeable visual and photographic objectives of 60-cm aperture, and with this instrument Loewy and V. Puiseux made their photographic atlas of the Moon with the aid of 10 000 plates, and thereby also advanced the study of the libration of the Moon.

In 1878 the 19-cm meridian circle constructed by Eichens was installed and presented to the

7.2. Paris Observatory, facing north, 1975. The view has been essentially unchanged since the turn of the century. In the right foreground is the dome of the Henrys' astrographic equatorial. On the main building the east dome contains the 36-cm Brunner refractor (1855) and the west dome contains the 30-cm Secrétan-Eichens refractor.

Observatory by the banker R. Bischoffsheim; this remained in service until 1960.

The *Carte du Ciel* was the principal concern of the Observatory during the directorships of F. Tisserand (from 1892 to 1896) and Loewy (from 1896 to 1907). Tisserand was noted for his *Traité de Mécanique Céleste* (1889–96), which remains to this day a fundamental work of reference. Loewy organized with the help of 55 observatories the programme of observations of the asteroid Eros, which passed close to the Earth in 1900–01; it was by analysis of the observations that A.R. Hinks of Cambridge derived the values of the parallax of the Sun and of the mass of the Moon, which were then universally adopted.

B. Baillaud, who was director from 1907 to 1926, played a decisive role in the setting up of the Bureau International de l'Heure. A regular service of time signals by radiotelegraphy was inaugurated on 23 May 1910; Commandant Ferrié, who was in charge of the telegraphic services at the Eiffel Tower, was responsible for transmitting the signals from an astronomical pendulum installed in the basement of the Observatory. At the instigation of the director, a Bureau International de l'Heure was created by international conferences called together in 1912 and 1913, and formally ratified in 1919, after World War I. Baillaud was likewise one of the founders of the International Astronomical Union formed in 1919, and he was elected the first President.

At the beginning of 1927 the Observatory of Meudon was united with that of Paris. The Meudon observatory had been set up in 1875 as a result of the efforts of P.J.C. Janssen to foster astrophysics in France; in Paris the site was unfavourable towards astrophysics and the director likewise. Janssen was granted the old royal estate of Meudon and authorized to spend over one million francs. This wooded region, 5 km southwest of Paris and dominating from 130 m the valley of the Seine, offered observing conditions that were excellent at the time and which have remained acceptable. Nearly twenty years were required to restore the structures remaining from the chateau dating from the time of Louis XIV. By 1893 the two principal instruments, built by Gautier with optics by the Henry brothers, were finally installed: a double refractor of 16-m focal length with objectives of 83 and 62 cm, and the reflector of 3-m focal length and 100-cm aperture. In the same year H. A. Deslandres at Paris built a spectroheliograph for the photography of the outer regions of the atmosphere of the Sun; installed in Meudon in 1898 and later improved, the instrument was enlarged to 15 m in 1909 and from that time provided daily information for research in solar physics.

Deslandres, who in 1908 had succeeded Janssen in charge of the Observatory of Meudon, was in 1927 appointed director of the combined establishments, which from that time were together known as the Paris Observatory.

Pulkovo Observatory
Aleksandr A. Mikhailov

The Pulkovo Observatory, founded in 1839, received its name from a small village to the south of Leningrad situated beneath the hill on which the Observatory was erected. The Observatory, described in detail in 1845 by its first director, F. G. W. Struve (1793–1864), was to be devoted to pure science, to practical geography and geodesy, and to the teaching and training of geodesists and hydrographers in practical astronomy. The original staff, which consisted of five astronomers including the director, soon proved inadequate for these obligations, and the years 1857–62 brought to Pulkovo a new set of statutes and a new director. In 1857 the staff was enlarged and the annual budget and salaries increased. Its personnel then comprised the director, four Senior Astronomers (one of them vice-director), two Adjunct Astronomers, two calculators, a mechanic, a janitor, a secretary and a physician. Besides these permanent personnel there was a chronometer-maker for the upkeep and winding of the time pieces, and a carpenter for the maintenance and repair of the numerous wooden structures.

For housing the increased staff and their families, the two wings of the main building were enlarged in 1859 by adding second stories. In 1856 a small building with two revolving domes 12 feet (3.7 m) in diameter was built for the work in geodesy. Another similar building with two domes 14 feet (4.3 m) in diameter was erected in 1866 and two portable transit instruments and a small Repsold vertical circle were acquired.

In 1862 the Observatory was formally transferred from the Academy of Sciences to the Ministry of Public Education. However, the connection with the Academy was retained through a special committee chaired by the president of the Academy and consisting of several scientists as well as representatives of related institutes. The Observatory director, who was elected by the Academy and thus became its full member, annually presented to the committee a detailed report on the activities, scientific achievements, staff and needs of the Observatory. The Senior Astronomers were also elected by the Academy on the recommendation of the director.

F. G. W. Struve remained director until 1861, although owing to his illness the Observatory was actually directed from 1858 by the vice-director, his son Otto W. Struve (1819–1905), who was formally elected director by the Academy in 1862. The new statutes of 1862 contained a detailed enumeration of the obligations and responsibilities of the director and members of the staff. The establishment of close connections with other Russian observatories was emphasized as well as the practice of regular visits to similar institutions in Russia and abroad. This latter obligation was fully realized; meanwhile, many outstanding foreign scientists visited Pulkovo and some even worked there.

In anticipation of its 50-year jubilee in 1889, the Observatory was equipped with several new instruments, including a 30-inch (76-cm) refractor, the largest in the world from 1885 until the erection of the Lick 36-inch (91-cm) in 1888. The lens was made by Alvan Clark in Cambridge, Massachusetts, and the mounting by A. Repsold in Hamburg (Figure 3.4). It was housed in a large new pavilion with a revolving dome 22 m in diameter. A museum of old astronomical instruments was installed in a lower circular gallery.

The development of astrophysics led to the establishment of a new position for an astrophysi-

7.3. Pulkovo Observatory, exterior view facing south, *c.* 1930. The large central dome housed the original 15-inch refractor. The east dome contained an Ertel transit and the west dome a Repsold meridian circle.

cist in 1882, and a separate astrophysical building with an electrical station was erected in 1886. This included a laboratory containing a large spectrograph with two Rowland gratings, a dark room for photographic work and several rooms for measuring apparatus and the processing of observations.

The library, having become too congested at its location in the anteroom of the building of the prime vertical transit instrument, was transferred to the more spacious mezzanine floor below the 15-inch (38-cm) refractor. An extensive library catalogue was published by O.W. Struve in two large volumes in 1860 and 1880.

The Observatory gained a great reputation for its work on stellar positions, and its determination of the constants of aberration, precession and nutation carried the largest weight when Simon Newcomb formulated new standard values for them in 1895. O.W. Struve himself became one of the leading double-star observers of his age, making 6080 micrometric measurements of 905 binary systems. As the number of observations increased, their reduction took up all the time of the staff, and so the previous large geodetic fieldwork programme had to be discontinued. However, the teaching and practical training of geodesists and hydrographers at the Observatory increased. Several expeditions were sent to observe the transits of Venus (in 1874 and 1882) and various solar eclipses, especially the one of 19

August 1887 in Spain. International scientific ties were strengthened, stimulated by the foundation of the Astronomische Gesellschaft in 1863, over which O.W. Struve presided from 1867 to 1878. In 1887 Struve was elected president of the International Astrographic Congress in Paris.

In 1889 O.W. Struve retired, and despite his desire to have his own son succeed him, F. Bredikhin (1831–1904) from Moscow was appointed director. He invited the spectroscopist A.A. Belopolsky (1854–1934), also from Moscow, to serve on the staff, and soon sent him abroad to order several new instruments, among them an astrograph of the *Carte du Ciel* type and a large stellar spectrograph to be attached to the 30-inch refractor. Belopolsky achieved an international reputation for his radial velocity measurements made with this instrument.

Bredikhin encouraged Russian astronomers to observe and work at Pulkovo and he himself visited most of the Russian observatories in order to stimulate cooperation with them. Because of ill health he retired in 1895; J.O. Backlund (1846–1916), who had worked at Pulkovo since 1879, was appointed his successor. One of Backlund's first acts was an application to the Ministry for permission to include women with college education on the permanent staff of the Observatory; however, this was turned down and women were allowed to work only temporarily, as piece-workers on a reduced salary.

During the period 1899 to 1901, in cooperation with Swedish scientists and several astronomers from Pulkovo, Backlund measured the northernmost meridian arc in Spitzbergen. Shortly after the discovery of the motion of the Earth's poles, a large zenith telescope was constructed in the mechanical shop of the Observatory and regular latitude observations by the Talcott method were begun in 1904.

As astrometrical observations of the Sun had to be interrupted in winter because of the low solar altitude at Pulkovo, a new transit instrument and vertical circle were acquired and installed at the more southern observatory of Odessa. In 1912 the observations at Odessa were transferred to the naval observatory at Nikolayev near the Black Sea, which became a branch of Pulkovo engaged mainly in astrometry and time service. In 1908 a private observatory at Simeis in the Crimea became affiliated with Pulkovo; at Simeis a photographic camera and a small refractor were used for observations of variable stars, minor planets, and comets, many of which were discovered there.

A 32-inch (81-cm) astrograph and 40-inch (102-cm) parabolic reflector were ordered from Grubb in Dublin. However, World War I delayed their manufacture, and the reflector was shipped to Simeis only in 1925. No suitable glass could be obtained for the lens of the astrograph so that only the mounting was made; sent to Pulkovo, it was destroyed during the siege of Leningrad in World War II.

Following the death of Backlund in 1916, Belopolsky became director. During the October Revolution of 1917 the Observatory passed through several anxious days when military action took place in its immediate neighbourhood. Life at Pulkovo became very difficult as a result of the shortage of food and absence of regular communication with the city, about twenty kilometres away. When conditions improved, temporary regulations for the administration of the Observatory were introduced. The director was to be elected for three years by a council of astronomers and confirmed by the Academy, and women could be members of the staff. In 1919 Professor A. Ivanov (1867–1939), then rector of the Petrograd University, was unanimously chosen as director. The staff was greatly enlarged and consisted of 22

astronomers, 12 computers, a librarian, four office clerks, an electrotechnician, and a mechanic with two assistants and three apprentices. The personnel of the Nikolayev and Simeis branches were similarly enlarged. After 1919, when fighting with bands of General Yudenich took place in the vicinity of Pulkovo leading to their defeat, the life of the Observatory began to return to normal.

During the next twenty years many innovations were introduced. In 1921 the Academy confirmed the new rules except that the director was to be elected for five years instead of three; Ivanov was re-elected in 1922. In the same year a large solar Littrow spectrograph with a coelostat arrived from England, and in 1924 Belopolsky began regular observations of the Sun's rotation. The time service was enlarged and the transmission of wireless time signals from Pulkovo organized. In 1924 a special committee chaired by the director was established to coordinate the time transmissions by the several observatories in the USSR and to regulate the standard time reckoning. In 1925 a new fundamental determination of longitude between Pulkovo and Greenwich was completed.

In 1927 Ivanov was once again re-elected director. He retired in 1931, and was succeeded by B. Gerasimovich from Kharkov. After Gerasimovich's death in 1937, S. Belyavsky from Simeis became director. In 1934 the Observatory was returned to the Academy of Sciences, thus becoming one of its oldest and most important institutes.

In 1926 the 40-inch (102-cm) reflector by Grubb was installed at Simeis and spectrographic determinations of radial velocities of stars begun. In 1927 a new astrograph by Zeiss for zonal observations was mounted in a special pavilion at Pulkovo and between 1929 and 1931 the photography of the northern polar zone was completed. Many smaller instruments were acquired, including those for observing solar eclipses, the latter being made in Leningrad and extensively used for observations of the eclipse of 19 June 1936. In 1940 a large horizontal solar telescope was constructed by a Leningrad optical factory to a design by the Pulkovo astronomer and engineer, N. Ponomarev, and mounted in a special building.

In the spring of 1940 the centenary of the Observatory was celebrated by the Academy, but this was almost the end of its existence: during

World War II, in August 1941, the Observatory was bombed and later came under direct artillery fire. All the buildings were totally destroyed. The instruments of medium size could be dismounted and hidden in a safer place but only the optical parts of the larger ones could be saved. Part of the famous library also perished. Several astronomers who could not be evacuated from the besieged city died from starvation.

After the war in 1946 the rebuilding of the Observatory on its historical site began under the directorship of A. Mikhailov, who was elected in 1947. An official opening of the restored and enlarged Observatory took place in May 1954.

Harvard College Observatory
Howard Plotkin

Except for a brief period between 1848 and 1855 when it was part of the newly-established Lawrence Scientific School, the Astronomical Observatory of Harvard College operated as a separate department of the University. The Corporation of Harvard University appointed both the director and his principal staff; each held an appropriate title: assistant, research associate, astronomer, or assistant professor. In addition, the Observatory statutes, passed in 1849, gave the director authority to add other members to the staff according to the needs and available income at any given time. These staff members had no official connection with the University since they were not Corporation appointees, and their number varied drastically, from one in 1865 to seventeen in 1888 (ten of whom were women). The director held the title of Phillips Professor of Astronomy from 1858 until 1887, when it was transferred to the chief assistant; from 1887 on, the director held the title of Paine Professor of Astronomy.

Although the Paine Professor, the Phillips Professor and the assistant professors were members of the Faculty of Arts and Sciences, they took no part in any University teaching, and no provision was ever made at the Observatory for the formal instruction of students. Since the statutes gave the directors authority over the Observatory's equipment and work (subject only to the superior authority of the Corporation), they used their authority to operate the Observatory as a purely research institution. Although several students served for brief periods as unpaid assistants,

thereby gaining invaluable practical experience, the University gave no official credit for such work.

During its first eighty years, the Observatory was administered by four directors: W.C. Bond (1789–1859), director from 1839 to 1859; his son G.P. Bond (1825–65), from 1860 to 1865; J. Winlock (1826–75), from 1866 to 1875; and E.C. Pickering (1846–1919) from 1877 to 1919. Between 1839 and 1844 W.C. Bond made his observations from a special cupola mounted on the roof of the Dana House. Using his own instruments – including a small transit, three small telescopes, two quadrants, two sextants and an astronomical clock – and receiving no salary from the University until 1846, Bond made observations of sunspots, transits, eclipses, occultations and comets. Most of his early observations, however, were meteorological and magnetic.

Harvard's permanent Observatory was erected in 1846 on a hill on Garden Street known as Summerhouse Hill, then on the outskirts of Cambridge. The 15-inch (38-cm) Great Refractor was mounted the following year; made by the German firm of Merz & Mahler, and paid for by funds subscribed by the citizens of Boston, the telescope was the largest in the United States at its time. Within months of its mounting, Bond used it in an attempt to obtain daguerreotypes of the Sun; by 1850 he and J.A. Whipple had daguerreotyped the Sun, Moon, planets and stars, the Harvard College Observatory being the first American observatory to apply this new technique to astronomy. These photographic investigations were continued and expanded under the directorship of G.P. Bond.

When Winlock became director in 1866, he ordered a new meridian circle from Troughton & Simms in England. When it finally arrived in 1870, he placed it under the direction of W.A. Rogers, an assistant professor, who used it in the international cooperative effort sponsored by the Astronomische Gesellschaft to revise Argelander's *Bonner Durchmusterung*. Winlock also continued the Observatory's experiments in photography, and began new researches in photometry, placing them under the direction of C.S. Peirce, then an assistant. Winlock led two solar eclipse expeditions, one to Shelbyville, Kentucky in 1869, and another to Jerez de la Frontera, Spain, the following year. A member of Winlock's party in Spain, C.A. Young of Dartmouth, discovered the flash

7.4. The plate stacks in Building C of the Harvard College Observatory, *c.* 1900.

spectrum and solar reversing layer at the 1870 eclipse.

Pickering's appointment initiated a period of immense growth in terms of the Observatory's staff, equipment, number and scale of investigations, and endowment. The Observatory received no money from the University. Its total income in 1877 was just over $14 000 a year (coming primarily from the sale of time service and publications but also including earnings from endowments that totalled roughly $175 000). Such an income, obviously, was scarcely enough to pay the salaries of the employees and cover the operating expenses of the Observatory. Pickering therefore found it imperative to find new sources of money, and, following the example of his predecessors, he launched a public appeal for funds. Within a year, the full amount he sought – $5000 annually for five years – was subscribed. He also included an

appeal for financial aid in his first Annual Report, and repeated it each succeeding year. Pickering himself often donated money to the Observatory, usually anonymously; in all, he gave more than $100 000.

With adequate support for the time being, Pickering launched his first major project, a quantitative determination of the magnitudes of all the stars visible from Cambridge. Although this photometric investigation continued for nearly a quarter of a century, with Pickering himself making more than 1.5 million readings, his 1883 Annual Report outlined far-reaching future plans that would transform the Observatory into the foremost institution of its kind in the United States: plans for a major study of the spectra of the stars, an augmentation of the Harvard glass plate collection of astronomical photographs which would thereby become a permanent history of celestial events duplicated nowhere else, and the first complete photographic map of the sky.

These plans, however, could not be carried out without a huge· increase in the Observatory's endowment. Partly through bequests, and partly as a result of Pickering's outstanding ability to secure money for the Observatory, Harvard obtained four major funds between 1886 and 1889 that enabled him to execute these projects. Two of these, the Robert Treat Paine Fund and the Henry Draper Fund, became available in 1886. Paine, a Boston lawyer and avid amateur astronomer, bequeathed his entire fortune amounting to more than $300000 to the Observatory. The bequest was used to support the Paine Professorship of Astronomy (held by the director) and for the routine operating expenses of the Observatory. Mrs Henry Draper, the widow of one of America's pioneering astrophysicists, yielded to Pickering's persuasive powers, and agreed to underwrite his long-range programme of photographing, measuring and classifying the spectra of the stars as a memorial to her husband. During her lifetime she gave more than $235000 to Harvard for this purpose, and bequeathed an additional $150000 on her death in 1914.

In 1887 a bequest amounting to $238000 from U.A. Boyden, a Boston engineer and mechanic, became available to Harvard. Since his will stipulated that the money be used to establish a high-altitude observatory, Pickering used it to build an auxiliary station at Arequipa, Peru at an elevation of about 2440 metres. Two years later, in 1889, Miss C.W. Bruce, a wealthy New York spinster, read one of Pickering's circulars in which he sought funds for the construction of a 24-inch (61-cm) photographic telescope to photograph and chart the entire sky. Miss Bruce responded with a $50000 gift to Harvard for that purpose. The Bruce telescope, ground and figured by the Clarks, was shipped to the Boyden station in Peru in 1895, where for more than thirty years it yielded important results (Figure 3.6).

By the late 1880s, therefore, Pickering had pieced together an impressive combination of Paine, Draper, Boyden and Bruce Funds. This was undoubtedly the high point of the Observatory's financial well-being during his administration. He referred to its condition at this time as a "kind of wealthy pauperism", for a fourfold increase in

endowment had brought with it a fivefold increase in expenses. Nevertheless, Pickering was now able to carry out his long-term and vast plans for measuring the magnitudes, classifying the spectra, photographing, and mapping the positions of the stars in both the northern and southern hemispheres.

Throughout the forty-two-year tenure of his administration, Pickering executed these plans with brilliant success. By combining celestial photography, introduced by W.C. Bond, with spectroscopy, begun by Winlock, Pickering brought together two techniques that became an essential feature of the Harvard College Observatory's research programme – and helped revolutionize astronomy in the process. In this manner, he transformed the Harvard College Observatory from a small, poorly endowed and equipped institution engaged in research in the 'old astronomy' of position and motion, into a large, well-endowed, well-equipped institution that became the leading American centre for research in the 'new astronomy', astrophysics.

United States Naval Observatory
Deborah Warner

Proposals for a national astronomical observatory were introduced into Congress as early as 1810, and repeatedly thereafter for several decades, with the support of such prominent politicians as Thomas Jefferson and John Quincy Adams, but to no avail. Despite a popular enthusiasm for science, most Americans were reluctant to commit public funds to support scientific research.

America's first national observatory was established surreptitiously, camouflaged by a practical necessity. The Department of the Navy set up a Depot of Charts and Instruments in Washington, DC in 1830, and authorized the celestial observations needed to rate the chronometers stored there. Once introduced, the astronomical activities increased rapidly. By 1842 Congress had authorized funds for a permanent building, for an astronomical observatory in fact, if not yet in name. The tension inherent in the Naval Observatory's inception remained throughout the nineteenth century, scientists viewing the Observatory as a basic research facility, and the Navy often stressing its support functions. In 1894 the objectives of the Observatory were defined as:

Primarily. *To determine accurately, from time to time, the positions of the Sun, Moon, planets, and stars, for use in preparing the Nautical Almanac . . .; to test, regulate, and perfect the character of chronometers . . .; to issue correct standard time daily to the public . . .; to fix accurately the longitude of various places by telegraph for geodetic and other surveys, and to furnish correct time, when requested, to other branches of Government and scientific departments; to investigate the subject of magnetism and meteorology as aids to navigation; and to distribute to vessels of the Navy instruments of precision for navigating purposes.*

Secondarily. *To conduct astronomical investigations of general or specific scientific interest, and to cooperate with other observatories, so far as time, opportunity, and the utilitarian character of the work required by the Department will admit.*

In 1838 the Depot of Charts and Instruments acquired a small meridian circle and a portable achromatic refractor. With these instruments Lieutenant J. M. Gilliss, in cooperation with the US Exploring Expedition to the South Seas, observed the Moon and Moon-culminating stars, solar and lunar eclipses, eclipses of Jupiter's satellites, and occultations of the brighter stars. Gilliss also determined the right ascensions of 1248 stars, and, in conjunction with the international 'magnetic crusade', made a series of observations of terrestrial magnetism. In 1842 he was assigned responsibility for planning the new observatory building and selecting the instruments.

Completed in 1844, the Naval Observatory was the most extravagant and best-equipped astronomical observatory in the United States. As there were as yet no indigenous instrument makers of proven ability, the original instruments were purchased in Europe. The selection was a mixture of British and German design and construction:

- achromatic equatorial telescope, 9.6 inches (24.4 cm) aperture, by Merz & Mahler of Munich.
- meridian transit, $5\frac{1}{3}$ inches (13.5 cm) aperture, by Ertel of Munich.
- prime vertical transit, 4.9 inches (12.4 cm) aperture, by Pistor & Martins of Berlin.
- mural circle, 5 ft (1.5 m) diameter, with telescope 4.1 inches (10.4 cm) aperture, by Troughton & Simms of London.
- comet seeker, 4 inches (10.2 cm) aperture, by Utzschneider & Fraunhofer of Munich.

- meridian circle, $4\frac{1}{2}$ inches (11.4 cm) aperture, by Ertel & Sohn of Munich.
- refraction circle, 7 inches (17.8 cm) aperture, by Ertel & Sohn of Munich.

Following English tradition, the mural circle was used for measuring declinations, and the transit instrument for right ascensions. German astronomers at that time were working with meridian circles, combining the two measurements with efficiency and accuracy, and in 1865 the Naval Observatory acquired from Pistor & Martins a first class, modern meridian circle of 8.52 inches (21.6 cm) aperture. In 1873 the Great Equatorial was installed, a 26-inch (66-cm) refractor made by Alvan Clark & Sons, the largest in the world at the time.

In 1893 the Observatory moved from its original location along the Potomac River, in an area known as Foggy Bottom. The new site, about three kilometres northwest, was on higher ground, further removed from city traffic, and spacious enough for separate buildings for the various instruments. Following the move the Observatory acquired two new instruments: a meridian circle of 6 inches (15 cm) aperture, and a 5-inch altazimuth, both designed by William Harkness of the Observatory staff, and both built by Warner & Swasey.

Scientific research at the Observatory related primarily to the determination of the fundamental celestial positions, motions and constants – the 'old astronomy' underlying the more glamorous problems of astrophysics and gravitational theory. An important contribution to the determination of planetary orbits was made in 1847, when Sears Cook Walker noticed that Neptune, seen by J.G. Galle just the year before, had been sighted by Lalande and listed twice in his star catalogue for 1795. Later, J. Ferguson discovered three asteroids, and in 1877 Asaph Hall discovered the satellites of Mars with the 26-inch refractor.

The Naval Observatory participated in the development and use of the so-called American method of determining longitude – comparing local time with that telegraphed from a master clock at a fixed observatory. The first signals for this purpose were sent between Washington and Baltimore in 1842. With a clock and chronograph made by John Locke, one of the pioneers of this method, the Naval Observatory exchanged time

7.5. US Naval Observatory in its old Washington location, *c.* 1885. The dome houses the 9.6-inch Merz & Mahler equatorial refractor.

signals with other observatories and with Coast Survey field parties. The Observatory's time service was initiated in 1865, and later extended to railroads across the nation and, via radio, to ships at sea. The Observatory sponsored expeditions to observe solar eclipses, and it served as headquarters for the commission coordinating the American expeditions to observe the transits of Venus of 1874 and 1882. Responsibilities for the *Nautical Almanac*, established by Congress in 1849, came to the Naval Observatory in 1893.

The scientific staff of the Naval Observatory has always included both civilians and Navy men, several of whom – notably Simon Newcomb, Hall and W. Harkness – made important contributions to astronomy. Newcomb served as a commissioned professor of mathematics at the Observatory from 1861 to 1877, when he was appointed Super-

intendent of the Nautical Almanac Office. Assigned first to the transit circle and later to the mural circle, he soon found that the Observatory routine allowed him sufficient free time to pursue independent research. Within a few years he ambitiously identified his research programme as that of developing improved theories and tables of planetary motion. Investigations of the motions of Uranus and Neptune, and later of the Moon, led to the publication of tables that were soon adopted for use in all of the major ephemerides of the world. He also played an instrumental role in securing the Great Equatorial for the Observatory, proposing its construction in 1868 and 1869 and negotiating the contract with the Clarks in 1870. Newcomb's inquiries dominated dynamical astronomy, and he became the best-known American scientist of his day.

By law, the Superintendent of the Observatory must be a naval officer. As it turned out, in the nineteenth century they changed command frequently – there were 13 between 1861 and 1900 – and several had no particular training in or enthusiasm for astronomy. Thus the first Superintendent (from 1844 to 1861), Lieutenant M.F. Maury, was mainly concerned with hydrography; the wind and current charts he developed significantly shortened ocean routes, and his *The Physical Geography of the Sea* was the first important book on that subject.

Despite the fact that the Naval Observatory has always been a branch of the Navy Department, it has often been viewed as *the* American national observatory. During the last quarter of the nineteenth century the professional astronomical community tried repeatedly to wrest control of the Observatory, to make it a national institution. In 1894 a compromise was reached in an attempt to lessen the tension between civilian scientists and bureaucratic command. The Naval Superintendent remained the commanding officer, but responsibility for the direction, scope, quality, quantity and publication of the scientific work was given to the Chief Astronomical Assistant, a man chosen for his scientific rather than military credentials.

In the first quarter of the twentieth century, the principal astronomical programme at the Observatory continued to be the making of observations of the Sun, Moon, planets and brighter stars; the determination of fundamental astronomical constants; the maintenance of its time service; and the preparation and publication of the *American Ephemeris and Nautical Almanac*. Expeditions were sponsored to observe the solar eclipses of 1900, 1901, 1905, 1918 and 1925; this last eclipse marked the first time that observations were successfully made from an aeroplane. Six asteroids were discovered between 1903 and 1924, and the Observatory served as a clearing-house for the publication of observations of the transit of Mercury of 1924. A photographic zenith tube, acquired in 1914, was used for determining the variation in latitude and the constant of aberration. By devoting itself to investigations in the 'old astronomy' at a time when most other observatories in the United States were devoting themselves to astrophysical researches, the Naval Observatory played a unique and important role in the development of American astronomy.

Lick Observatory
Trudy E. Bell

The Lick Observatory on Mt Hamilton, California, was made possible by the Deed of Trust of the eccentric piano-maker and landholder James Lick (1796–1876), who before his death set aside $700000 – nearly a quarter of his estate of three million dollars – to build "a powerful telescope, superior to and more powerful than any telescope ever yet made . . . and also a suitable observatory connected therewith". Lick originally intended the completed observatory to be turned over to the State of California, but he was soon persuaded to give it instead to the newly-founded University of California. The title to the observatory was formally transferred from the Lick Board of Trustees to the Regents of the University of California on 1 June 1888.

Although the observatory was called the Lick Astronomical Department of the University of California and the staff was appointed by the Regents at the request of the observatory's director, it did not undertake the formal instruction of undergraduates, and graduate instruction was restricted to research students qualified enough to be functioning astronomical assistants. Moreover, the astronomers were not members of the Academic Senate (which included all persons giving instruction at the University of California), and residence at the observatory on Mt Hamilton was not considered residence at the University of California.

The site on the top of Mt Hamilton, twenty-one air kilometres east of San Jose, was chosen by Lick himself, with encouragement from G. Davidson, President of the California Academy of Sciences, who had advised him to locate the observatory on top of a mountain. The Lick Observatory thereby became the first major research observatory to be located on a mountaintop, and its early example and superior results encouraged subsequent American and foreign astronomers to establish observatories on mountain peaks.

As a site for the Lick Observatory 1350 acres (546 hectares) were granted by an Act of Congress; later land purchased or granted by the California State Legislature and by Congress increased the

Observatory's holdings to 4581 acres (1854 hectares). The site's isolation demanded that it be completely transformed from dusty fire-prone chaparral to a livable area for a self-sufficient community complete with post office, homes and a school for the astronomers' families. The construction of the Observatory, under the enthusiastic and dedicated supervision of R.F. Floyd, President of the Lick Trust, took well over a decade. In 1876 the County of Santa Clara built a road to the summit, and in 1880 the peak of the mountain was levelled and the buildings begun with bricks made from native mountain clay.

The Observatory's main telescope – the great 36-inch (91-cm) refractor built by Alvan Clark & Sons – was not installed until 1888, this marking the formal completion of the Observatory. Nonetheless, other instruments were mounted and some research begun about seven years earlier. In 1881 the 4-inch (10-cm) combined transit and zenith telescope by Fauth & Co. of Washington, DC was mounted in the transit house and the 12-inch (30-cm) Clark refractor was installed in its dome; both instruments were in operation in time to observe the transit of Mercury late that year. In 1882, about three weeks before the transit of Venus in December, the Clark horizontal photoheliograph was mounted. In 1884 the 6-inch (15-cm) Repsold meridian circle was delivered and installed.

From its inception until 1930, the Lick Observatory was administered by three directors: E.S. Holden (1846–1914), director from 1874 to 1897; J.E. Keeler (1857–1900), from 1898 to 1900; and W.W. Campbell (1862–1938), from 1901 to 1930. Holden acted as advisor while the Observatory was being built and he selected the original staff. He had been assistant at the US Naval Observatory to Simon Newcomb, chief astronomical advisor to the Lick Trust, and was then director of the Washburn Observatory at Madison, Wisconsin, from 1881 to 1885. Holden became President of the University of California in 1886, but resigned two years later to become director of the Observatory when the newly-completed institution was transferred from the Trustees to the University.

The construction of the original observatory cost $610000 (of which the instruments accounted for about $112000); the remaining $90000 from James Lick's Trust was invested as an endowment, the income from which was used to sup-

plement the yearly support from the State of California. Together, however, the two sources of funds provided an income of only $25000 per year, only enough to maintain some six or seven astronomers (a fact that continually chafed Holden, who never lost an opportunity to point out that comparable institutions – such as the US Naval Observatory, the Harvard College Observatory and the Royal Greenwich Observatory – had between two and seven times the staff, and triple the income). Because of the financial struggle for maintenance itself, not to mention astronomical expeditions, new equipment, fellowships and special projects, the Lick astronomers were forced to rely on private philanthropy. Holden inspired donors to provide for various expeditions and instruments, his biggest achievement being the acquisition of the 36-inch reflector and dome owned by E. Crossley of Halifax, England. When the Crossley reflector was installed on Mt Hamilton in 1896, it was the largest reflecting telescope in America. In addition, between 1889 and 1918 some dozen total solar eclipse expeditions to various parts of the world were supported by the wealthy banker C.F. Crocker; between 1902 and 1905 a new mount for the Crossley reflector was built by means of donations from Mrs Phoebe A. Hearst; and over a period of four decades various instruments were purchased through the faithful generosity of D.O. Mills.

Although Holden undertook research himself, his principal genius lay in organizing and in writing. With an eye to the future of astronomy as well as to its present, for his first staff he recruited four talented astronomers whose collective expertise spanned both traditional positional astronomy and the 'new astronomy' of astrophysics: S.W. Burnham (1838–1921), a double-star observer; J.M. Schaeberle (1857–1923), a meridian-circle observer; E.E. Barnard (1857–1923) a deep-sky photographer and comet-observer; and Keeler, a spectroscopist. The fact that ultimately he was unable to hold his staff together tells as much about the lonely conditions on an isolated mountain, especially difficult for families, as it does concerning any failure in Holden's own personality. Nevertheless, antagonism to Holden was part of the reasons for Burnham's and Barnard's departures.

Keeler left the Lick Observatory in 1891 to

7.6. Lick Observatory, Mt Hamilton, California, *c.* 1910. The principal dome houses the 36-inch refractor, and the astronomers' residence stands in the foreground just below the main observatory building.

become director of the Allegheny Observatory near Pittsburgh, Pennsylvania. Following Holden's resignation in 1897, Keeler was appointed director of the Lick Observatory. Much more research-oriented than his predecessor, Keeler worked extensively with the Crossley reflector, photographing tens of thousands of spiral nebulae and measuring the radial velocities of 400 bright stars.

Shortly after Keeler's premature death in 1900, Campbell was appointed director. Campbell had come to Lick Observatory in 1890 to assist Keeler in his spectroscopic work, and upon Keeler's departure to Allegheny in 1891 was appointed astronomer in charge of spectroscopic observations. By the time he became director a decade later, Lick Observatory's work in both traditional astronomy and astrophysics was significant: Lick astronomers had accurately measured the positions of thousands of stars, had discovered the fifth satellite of Jupiter, 25 comets, and 1300 new double stars (including 40 spectroscopic binaries); in addition, they had photographically recorded for the first time the structure of the Sun's inner corona, and had analysed the composition of gaseous nebulae,

peculiar stars, novae and the atmosphere of Mars. Under Campbell the work continued, but his own principal contribution was in organizing and carrying through an extensive stellar radial-velocity programme. In 1922, during a total solar eclipse in Western Australia, he obtained the observations to confirm the marginal results of A. S. Eddington and his associates who in 1919 had found the gravitational deflection of starlight predicted by Einstein's general theory of relativity. Upon returning from the expedition, Campbell was offered the presidency of the University of California. As the University's president from 1923 to 1930, Campbell nominally retained the title of Director of the Lick Observatory (without remuneration) but the day-to-day affairs of the observatory were handled by his associate director R. G. Aitken.

When Campbell first became director, he planned to issue a complete catalogue of stellar radial velocities, which required extending the work to include stars in the southern hemisphere. Mills agreed to provide funds to erect and maintain a southern observing station of the Lick Observatory. In 1904 the observatory of the D. O. Mills

Expedition was established on Cerro San Cristóbal northeast of Santiago, Chile, with a 37-inch (94-cm) Cassegrain reflector built by John A. Brashear & Co. With that telescope, the spectra of all the bright stars south of declination -25 degrees were photographed. Although the original plans were to operate the observing station for only a couple of years, the importance of the work induced Mills to support it for the rest of his life and beyond. It was operated continuously by the Lick Observatory until 1929, when it was sold to the Catholic University of Santiago.

The Lick Observatory, a great institution founded by one man's vanity on the crest of the wave of observatory-building in nineteenth-century America, was a pioneer of the new observatories built with an eye explicitly on the needs of future generations: located in a remote site, emphasizing the fledgling 'new astronomy'. Its continued success up until the present attests to the wisdom and vision of its early leadership.

Potsdam Astrophysical Observatory
Dieter B. Herrmann

The first suggestion for the founding of a special institute in Germany for the investigation of the Sun came from the noted sunspot observer, G. Spörer, who was at that time a professor in the secondary school at Anklam. In 1871, at the instigation of K. H. Schellbach, the former teacher of the German Crown Prince, a "Memorandum concerning the establishment of a solar observatory" was drawn up by W. Foerster, the influential director of the Royal Observatory in Berlin. Foerster stressed both the great practical significance of solar research, due to the influence of the Sun upon the Earth, and the pre-eminence that German science might attain through the creation of such an institute, but he omitted any reference to the spectral analysis of the Sun, in which the Heidelberg physicist G. R. Kirchhoff already excelled.

The next step in the eventual establishment of the Potsdam Astrophysical Observatory was to seek the opinion of the permanent astronomical commission of the Royal Academy of Sciences. After a series of discussions the project was finally recommended, although members of the Academy were of the opinion that the scope of research should include the whole of the recently-emerged

subject of astrophyscs, the 'new astronomy'. In 1873 a special temporary commission of the Academy met, with H. v. Helmholtz, A. Auwers and E. Dubois-Reymond among its members, and took the project forward to the stage where a cost plan could be drawn up.

The institute commenced activity in July 1874, with a team consisting of Spörer and H. C. Vogel (1841–1907) as observers and Oswald Lohse as assistant. Vogel had already begun his researches in stellar spectroscopy at Bothkamp, a private observatory near Kiel, and Lohse had been his photographic assistant; at Potsdam Lohse promptly installed photographic workrooms and laboratories. Spörer carried on the sunspot measurements that enabled him to establish more accurately the differential rotation of the Sun, working at first from a tower in Potsdam.

The Academy commission had hoped that Kirchhoff, whose research had played a major role in the creation of astrophysics, would become director of the new observatory, but he refused on the ground that the post would leave him insufficient time for his studies in theoretical physics. He agreed, however, to move to Berlin and to serve on a board of co-directors, along with Foerster and Auwers. As a Member-in-ordinary of the Academy, Auwers took over the administrative responsibilities, an arrangement that continued until Vogel was appointed director in 1882.

The year 1876 saw the building of the Observatory on Telegraphenberg, a vibration-free hill in Potsdam about 30 kilometres southwest of the centre of Berlin. The main buildings included three observing domes and several laboratories and workshops. Initially the instrumentation consisted of a refractor by A. Repsold with a 30-cm f/18 objective by H. Schröder, a refractor by Grubb (20 cm), a Steinheil 13-cm telescope and a 16-cm heliograph with optics by Schröder. In 1889 a double refractor (visual objective 24 cm, photographic objective 32 cm by Steinheil and Repsold) was brought into service for making a survey of the $+32°$ to $+39°$ zone of the *Carte du Ciel*. In 1899 a large double refractor by Steinheil and Repsold (80-cm f/15 photographic objective, 50-cm visual objective) was completed, which remains to this day the world's fourth largest refractor (Figure 3.1); with this Vogel could extend his spectroscopic investigations to weaker stars.

7.7. Potsdam Observatory, *c.* 1930. The large dome houses the 80-cm refractor, and the tower of the geodetical institute appears just beyond the main building.

From its foundation until the end of Vogel's directorship in 1907 – and later, as well – the Observatory produced pioneering work in the field of astrophysics. Here visual work at first played a dominant role, for it was only later that photographic investigations achieved pre-eminence. Special mention should be made of the photometric survey of *BD* stars up to magnitude 7.5 in the northern sky ("The Potsdam Survey", by G. Müller and P. Kempf, 1886–1906) and also the spectroscopic survey of stars up to magnitude 7.5 from δ −1° to 20° (Vogel and Müller, 1880–82). Photographic work included the *Carte du Ciel* (J. Scheiner, O. Birck and others, 1883–1921), spectral investigations and the determinations of radial velocities (Vogel and others, from 1882), and the classification of the spectra of 528 stars in the northern sky (Vogel and J. Wilsing, 1895–98). The spectrophotometric work of Wilsing and Scheiner

(1909) was among the first to use Planck's law for interpreting colour temperatures.

Vogel died in 1907, leaving a large endowment to the Observatory for study abroad and for the support of gifted children. Late in 1909 K. Schwarzschild (1873–1916) came from Göttingen to become director of the Observatory. Making the move with him was E. Hertzsprung, who spent the next decade at Potsdam; there he extended his photographic and photometric studies, and in 1913 he determined the distance to the Small Magellanic Cloud from its Cepheid variables.

At Potsdam Schwarzschild continued his interest in the theory of radiative transfer in stellar atmospheres, and he began spectrographic researches, including a plan to derive stellar radial velocities using a reversible objective prism, and an unsuccessful attempt to observe the gravitational red shift in the solar spectrum. In 1910

Schwarzschild attended the fourth meeting of the International Union for Co-operation in Solar Research at Mt Wilson, and *en route* he visited the major American observatories. Schwarzschild was particularly impressed by the idea of the southern stations operated by Harvard and by Lick, and upon his return he began to push (without success) for a German observatory south of the equator.

In 1914 Schwarzschild volunteered for military service, first directing a weather station in Belgium, and afterwards calculating trajectories for long-range missiles in France and then on the Eastern Front. From Russia he communicated his famous papers on general relativity and on what is now called the 'Schwarzschild radius'. While there he contracted a rare and incurable skin disease, which led to his death in Potsdam in 1916 aged 42.

Schwarzschild was succeeded by the Potsdam photometrist G. Müller (1851–1925), who served as director from 1917 to 1921. During this time the Einstein Foundation was created and construction was begun on the Einstein Tower for solar physics. Completed in 1925, the tower telescope with its 60-cm Zeiss objective was designed to continue the studies on the solar gravitational red shift begun by Schwarzschild. During this same period (1923–25) new physical and spectrographic laboratories were added, and the staff was enlarged to include young astrophysicists such as W. Grotrian, E. von der Pahlen, E. Freundlich and H. von Klüber. These men plus numerous guest investigators, under H. Ludendorff (1873–1941) as Observatory director, maintained Potsdam's reputation as the leading German observatory of the twentieth century.

Further reading

Introduction

C. André and G. Rayet, *L'Astronomie Practique et les Observatoires* (Paris, 1874–78)

J. L. E. Dreyer, Observatory, in *The Encyclopaedia Britannica* (11th ed., Cambridge, 1911), vol. 19, 953–61

In addition to these references, many of the observatories have published their own histories, and of these we may mention F. J. M. Stratton, The history of the Cambridge observatories, *Annals of the Solar Physics Observatory, Cambridge*, vol. 1 (1949), and B. Boss, *History of the Dudley Observatory* (Albany, 1968); others are listed in the following sections.

Greenwich Observatory

The report of the Astronomer Royal for the Board of Visitors, *Greenwich Observations* (1860 to 1920 inclusive)

Derek Howse, *Greenwich Observatory*, vol. 3: *The Buildings and Instruments* (London, 1975)

W. H. McCrea, *Royal Greenwich Observatory* (London, 1975)

A. J. Meadows, *Greenwich Observatory*, vol. 2: *Recent History (1836–1975)* (London, 1975)

Paris Observatory

Rapport annuel de l'Observatoire de Paris pour 1878 (Paris, 1879) and succeeding reports up to and including 1927

G. Bigourdan, Le Bureau des Longitudes, son histoire et ses travaux [from 1795 to 1874], in *Annuaire du Bureau des Longitudes pour 1928* and continuing in the next five years

P. Levert, F. Lamotte and M. Lantier, *Le Verrier* (Coutances, France, 1977)

R. Mouchez, *Admiral Mouchez* (Paris, 1970)

Pulkovo Observatory

Pulkovskoi observatorii 125 let: sbornik statei (*Pulkovo Observatory's 125 Years: a collection of articles*) (Moscow-Leningrad, 1966)

Sto let Pulkovskoi observatorii: sbornik statei (*A Century of the Pulkovo Observatory: a collection of articles*) (Moscow, 1945)

Zum 50-jährigen Bestehen der Nicolai-Hauptsternwarte (St Petersburg, 1889; also in Russian)

A. N. Dadaev, *Pulkovskaya Observatoriya* (Leningrad, 1972) (an English translation by Kevin Krisciunas, *Pulkovo Observatory*, NASA Technical Memorandum TM-75083, is available from National Technical Information Service, Springfield, Virginia, USA)

Harvard College Observatory

Solon I. Bailey, *The History and Work of Harvard Observatory, 1839 to 1927* (New York and London, 1931)

Bessie Zaban Jones and Lyle Gifford Boyd, *The Harvard College Observatory* (Cambridge, Mass., 1971)

Howard Plotkin, Edward C. Pickering and the endowment of scientific research in America, 1877–1918, *Isis*, vol. 69 (1978), 44–57

United States Naval Observatory

Arthur L. Norberg, Simon Newcomb's early astronomical career, *Isis*, vol. 69 (1978), 209–25

J. E. Nourse, Memoir of the founding and progress of the United States Naval Observatory, *Washington Observations for 1871* (Washington, D.C., 1873), Appendix IV

Howard Plotkin, Astronomers versus the Navy: The revolt of American astronomers over the management of the United States Naval Observatory, 1877–1902, *Proceedings of the American Philosophical Society*, vol. 122 (1978), 385–99

A. N. Skinner, The United States Naval Observatory, *Science*, vol. 9 (1899), 1–16

Gustavus A. Weber, *The Naval Observatory, Its History, Activities and Organizations* (Baltimore, 1926)

Lick Observatory

Rosemary Lick, *The Generous Miser* (Los Angeles, 1967)

F.J. Neubauer, A short history of the Lick Observatory, *Popular Astronomy*, vol. 58 (1950), 201–22, 318–34, 369–88

Simon Newcomb, The Lick Observatory, chap. 7 in *The Reminiscences of an Astronomer* (London and New York, 1903)

D.E. Osterbrock, The rise and fall of E.S. Holden, *Journal for the History of Astronomy*, vol. 15 (1984), 81–127, 151–76

D.E. Osterbrock, *James E. Keeler, Pioneering American Astrophysicist* (Cambridge, 1984)

Potsdam Astrophysical Observatory

W. Hassenstein, Das Astrophysikalische Observatorium Potsdam in den Jahren 1875–1939, *Mitteilungen des Astrophysikalischen Observatoriums Potsdam*, no. 1 (1941)

D.B. Herrmann, Zur Vorgeschichte des Astrophysikalischen Observatorium Potsdam (1865 bis 1874), *Astronomische Nachrichten*, vol. 296 (1975), 245–59

J. Wempe, Zum 100. Jahrestag der Gründung des Astrophysikalischen Observatoriums Potsdam, *Die Sterne*, vol. 51 (1975), 193–206

8

Building large telescopes, 1900–1950
ALBERT VAN HELDEN

Introduction

If the second half of the nineteenth century was the golden age of the refractor (see Chapter 3), in the twentieth century the reflector has supplanted the refractor as the primary instrument of the astronomer. Indeed, modern astronomers consider the refractor obsolete except for a few special purposes. The dominance of the great reflector began when the limit of aperture of the refractor had been reached, and it paralleled the rapid growth of astrophysics, a science that made different demands on the telescope.

The great reflector became the central instrument in the new *astrophysical* observatory, of which the Mt Wilson (Solar) Observatory was the prototype. Here, under the guidance of George Ellery Hale (1868–1938), the funding problems presented by the new scale of science were largely solved, three generations of large reflectors pioneered, and the modern pattern of staffing and instrument use developed.

The advantages and disadvantages of reflectors

In the continuing quest for greater light-gathering power, the refractor was rapidly approaching its limit at the turn of the twentieth century. The manufacture of discs of crown and flint glass of optical quality with diameters greater than 100 cm was a formidable and prohibitively expensive undertaking with a low rate of success. Manufacturers of 'plate-glass' could, however, cast large discs of ordinary crown glass of lower optical quality, but suitable for mirrors, easily and cheaply. Moreover, calculations showed that light-losses in objectives of more than 100-cm aperture would be greater than those in mirrors of equal aperture, especially in the blue and violet region of the spectrum where the highest sensitivity of

photographic plates lay. There were also persistent questions as to the possibility of supporting larger objectives in their cells without flexing due to their great weights; a mirror could be supported at the back.

Whereas for visual purposes the residual chromatic aberration of a crown–flint objective was no serious handicap, for photographic and spectroscopic purposes it was. A photographic corrector was objectionable because it was awkward and expensive, and it changed the focus and increased light-losses. Spectroscopic work became exceedingly difficult if the different parts of a spectrum could not be brought to a focus in the same plane. The astrophysicists wanted the largest possible aperture and perfect achromatism, and only a concave mirror as primary receptor could give them that. Acceptable designs for reflectors also allowed smaller f-ratios than large refractors, a property that made reflectors preferable for stellar and nebular photography, for spectroscopy and for bolometric investigations.

The reflector's low f-ratio meant that for a given aperture it had a shorter tube than the refractor. This made observing easier and allowed smaller domes, which meant much lower construction costs. The tube itself could consist of an open framework, thus making it lighter, and since the mirror was supported at its bottom, the telescope was easier to balance. These factors, added to the much smaller cost of the primary receptor, made the reflector a much cheaper instrument than the refractor.

Nevertheless, around the turn of the century there was still considerable opposition to large reflectors among observatory astronomers who, for good reasons, considered these instruments to be imprecise and awkward. Observers used to the refractor found it difficult to tolerate the slight image distortion caused by the flexing of the mirror, and they pointed to the mirror's sensitivity to temperature changes as a major drawback of the reflector. Moreover, there were few, if any, reflectors whose mountings were well enough engineered to meet the demands of observatory work. The reflector was, therefore, still seen as an instrument of amateurs, and it took some time for it to emerge from the shadow of the refractor at major observatories. Thus, at Greenwich in 1897, a 30-inch (76-cm) Common reflector was mounted as a counter-weight on the other side of the declination axis of the Thomson photographic refractor and acted in a supporting role. At Lick, with its superior seeing conditions, J.E. Keeler (1857–1900) spent the two years before his untimely death modifying the awkward 36-inch (91-cm) Crossley reflector (also made by Common) and proving to a reluctant staff that important photographic work on nebulae – impossible with the 36-inch Lick refractor – could be done with it. But not until this telescope had been completely reconstructed under C.D. Perrine, between 1902 and 1905, did it become an instrument of acceptable precision and convenience. Perrine completed Keeler's programme of photography and the results were published in 1908. These photographs were a crucial victory for the reflector. But if the Crossley reflector proved the optical superiority of reflectors for certain applications, the important breakthrough in mechanical feasibility came, ironically, at the Yerkes Observatory, the site of the world's largest refractor.

Hale and Yerkes

From an early age, Hale was an enthusiastic advocate of the young science of astrophysics. During his student days at MIT he had built the first successful spectroheliograph, and when upon his graduation in 1890 he built his private Kenwood Observatory in Chicago, he devoted it entirely to the astrophysical study of the Sun. In 1892 he accepted an invitation to become Associate Professor of Astral Physics and Director of the Observatory of the new University of Chicago, and in that capacity he was largely responsible for the foundation of the Yerkes Observatory. He persuaded the streetcar magnate C.T. Yerkes to donate the money; he designed the observatory, supervised its construction and staffed it. Besides

being a very fine scientist, Hale was a superb organizer and money raiser, driven to undertake project after project. In his lifetime he changed the institutional structure of the observatory, put his personal stamp on the profession of astrophysics, reformed the National Academy of Sciences and pioneered international cooperation in science. He also transformed the Throop Institute in Pasadena, a glorified high school, into the California Institute of Technology, and was instrumental in the foundation of the Huntington Library and Art Gallery, an important institution for research in the humanities. For these efforts he paid dearly. After about 1910, his deteriorating health forced him to progressively curtail his manifold activities.

Whereas most of the older generation of professional astronomers were not oriented towards physics and still saw the observatory solely as a place for observation and calculation, Hale and his astrophysical colleagues saw it also as a laboratory: the function of the telescope was to provide light to other instruments for recording and analysis. As its director, Hale shaped the Yerkes Observatory accordingly. Between 1897 and 1903 he proved the superiority of the reflector for spectroscopy and most photographic purposes, made great strides in the instrumentation for solar studies, and showed that an instrument shop, where design and fabrication of ancillary instruments is a constant activity, is an indispensable part of the observatory.

Even as Hale was supervising the installation of the 40-inch (102-cm) refractor (see Chapter 3), he was advocating the cause of the reflector. As early as 1894 his father agreed to help in securing a large reflector for the Yerkes Observatory, and a 60-inch (152-cm) glass disc was ordered from the St Gobain glass works in France, at a cost of $25 000. In 1896, for the purpose of figuring the disc, Hale appointed to the Yerkes staff G.W. Ritchey (1864–1945), a young optician who had made several excellent reflectors of modest aperture. Ritchey had done some preliminary work on the disc when, in 1898, the death of Hale's father ended all hope of securing funds for mounting the instrument, and Ritchey therefore stopped his efforts. Instead, he decided to concentrate on a more modest reflector for photographic purposes. At his own expense, Ritchey made a 24-inch (61-cm) silvered-glass mirror with a focal length of 236 cm. The mirror

May 6 1921

8.1. The 40-inch Yerkes refractor during a visit from Einstein, 6 May 1921.

was housed in a skeleton tube of eight steel pipes connected by aluminium rings. The instrument was mounted as a German equatorial, and it proved sufficiently rigid to allow long exposures. Photographs taken at its prime focus were excellent and compared very favourably with the best photographs obtained with the 40-inch (f/18.5) refractor (which Ritchey had in the meantime equipped with a yellow filter for photography). In the *Carnegie Institution Yearbook* in 1902 Hale wrote: "With this small instrument stars too faint to be seen or photographed with the 40-inch Yerkes telescope (the largest refractor hitherto constructed) can be photographed in forty minutes." Because of its novel mechanical features, Ritchey's reflector drew the attention of the astronomical community, and these features were copied in other instruments.

In the meantime, in an effort to take advantage of the 18-cm solar image of the great Yerkes refractor, Hale had built a large spectroheliograph. But attaching this 317-kg instrument to the refractor, adjusting it, and counterbalancing its weight, was a time-consuming task that precluded any other instrument being attached to the refractor that day. Hale therefore sought a more convenient and flexible way to study the Sun. He and Ritchey built a long-focus horizontal reflecting telescope into which the Sun's image was introduced by a 30-inch (76-cm) coelostat mirror. But in December 1902, while it was still in the testing stage, this instrument was completely destroyed by fire. A gift from Miss H. E. Snow enabled Hale and Ritchey to build a new instrument of improved design. The Snow telescope, finished in 1903, was a limited success: although the solar image could con-

8.2. George Ritchey's 24-inch reflector, Yerkes, *c.* 1900. The excellence of its optics and mounting helped pave the way for the larger photographic reflectors.

veniently be introduced into several instruments in quick succession, definition of the image left something to be desired. One problem, potentially amenable to a solution, was mirror distortion due to solar heating. A problem not solvable at Yerkes was the poor seeing, which affected the solar image severely. Hale felt that perhaps the Snow telescope would perform better in the dry, stable air on a mountain top.

The founding of the Mt Wilson Solar Observatory

Although the Yerkes Observatory had quickly become a leading centre for astrophysics and attracted numerous foreign visitors, its financial resources were very limited. To acquire additional instruments Hale had to solicit money from other donors. The staff was inadequate and poorly paid, and the 60-inch (152-cm) mirror disc languished in the basement. The struggling University of Chicago could do little to help. What was needed for such large research establishments was a new form of funding, not directly dependent on the whims of uninformed (though often well-meaning) donors such as Lick and Yerkes. The newly-established Carnegie Institution of Washington, founded in 1902, held out much hope.

Since 1902 Hale had been a member of the

Institution's Advisory Committee on Astronomy, but his ambitious plan for astrophysics was only one of many projects deemed important by the committee members. By 1903, however, a small grant allowed W.J. Hussey of the Lick Observatory to search for a suitable site for a possible solar observatory. Hussey's glowing report on the seeing conditions at Mt Wilson, overlooking Pasadena in southern California, brought Hale there in the summer of 1903. Overjoyed by the seeing conditions on the mountain, Hale took a huge personal risk and moved his family and a solar telescope to Mt Wilson and began building an observatory, while the Advisory Committee on Astronomy and the Executive Committee of the Carnegie Institution debated the merits of a solar observatory on the mountain. Finally, in December 1904, the Executive Committee voted an appropriation of $300000 for the foundation of the Mt Wilson Solar Observatory. Hale became its first director.

The Mt Wilson Solar Observatory (after 1920 the Mt Wilson Observatory) set the style for modern observatories. Whereas Yerkes had still been built on the traditional pattern of an imposing central building with domes for the various telescopes and had houses for the staff and their families on the observatory's grounds, at Mt Wilson itself Hale built only the observing instruments and a 'monastery' with temporary accommodation for the staff actually engaged in observing at any one time. The staff lived in Pasadena where the Observatory's headquarters was located. Here the astronomers had their offices with equipment for the measurement of spectograms and direct photographs. They were assisted by a computational division with a growing membership. The Pasadena facilities included a photographic laboratory, a design department, an instrument shop and an optical shop. There was even a facility for making diffraction gratings. In these laboratories and shops new instruments were continuously developed. For the first fifty years of its existence, the Mt Wilson Observatory was the only observatory to have a well-staffed laboratory for conducting fundamental investigations in spectroscopy and radiation.

The problems that had traditionally plagued astronomers in obtaining funds for new instruments and, perhaps even more importantly, for

maintaining them and staffing the observatory, were substantially eliminated in the case of Mt Wilson. The Solar Observatory was an operating department of the Carnegie Institution. Informed budgetary decisions on staffing and maintenance provided for continuity and full use of the facilities. Ambitious expansion projects, when properly justified, were funded whenever possible with intelligent allowances for concomitant increases in staff. The initial appropriation of $300000 was spent in two years, after which funding was maintained at about $100000 per year until 1911, when it was more than doubled. By 1915 the Observatory had received about $1500000, of which about $1000000 had been for permanent installations, the remainder being for operating expenses.

From Yerkes Hale brought with him Ritchey together with F. Ellerman (1869–1940), W.S. Adams (1876–1938) and F.G. Pease (1881–1938). Ellerman, who had been his personal assistant since 1892 at Kenwood, continued in that role. The astronomer Adams became second in command and was acting director during Hale's frequent illnesses and business trips; he succeeded Hale as director in 1923. Ritchey was in charge of construction during the initial phase and continued as chief optician until his abrupt departure in 1919, while Pease concentrated on instrument design. The staff expanded with time. In 1908 A.S. King (1876–1957) came to Mt Wilson to head the physical laboratory where he was joined by H.D. Babcock (1882–1968) in 1909 and J.A. Anderson (1876–1959) in 1916. Adriaan Van Maanen (1884–1946) joined the astronomical staff in 1912; Harlow Shapley (1885–1972) came in 1914 and Edwin P. Hubble (1889–1953) in 1919. An increasing number of women were hired for the computing division, first headed by Adams and from 1909 by F.H. Seares (1873–1964). Instrument time was scheduled well in advance, so that the instruments saw maximum use. Observers had comparative freedom to explore areas of interest, although a rough division of labour was instituted and maintained.

Because of its concentration of instruments and astrophysicists, and its excellent support facilities, Mt Wilson quickly became the Mecca of astrophysics. Visitors from all over the world came regularly, especially during the summer, and Hale encouraged this practice by helping many of them secure funds from the Carnegie Institution through its Research Associate programme. J.C. Kapteyn from the Netherlands was the most frequent summer guest during the early years. Other foreign astronomers included G. Abetti (Italy), H. Chrétien (France), E. Hertzsprung (Germany), K. Lundmark and B. Lindblad (Sweden), H.H. Turner (England) and J.H. Oort (Netherlands). Among the Americans were E.E. Barnard (Lick), A.A. Michelson (University of Chicago), Henry Norris Russell (Princeton) and Joel Stebbins (Wisconsin).

Solar telescopes and the 60-inch reflector

Even before the foundation of the Observatory, Hale had obtained permission to move the Snow telescope from Yerkes to Mt Wilson, and by 1905 the instrument had been erected at its new home. Its 76-cm coelostat and a flat mirror were mounted on a pier built up from the mountain side. These mirrors cast the Sun's rays on one of two 61-cm concave mirrors with focal lengths of 18.3 and 44.2 m, which produced solar images of 18 and 41 cm respectively. A number of spectrographs and spectroheliographs could be wheeled into position to receive the solar image. Although much important work was done with the Snow telescope, mirror distortion and air turbulence remained problems and Hale therefore designed a vertical instrument in the hope of eliminating these. The new instrument employed a 60-ft (18-m) tower to house a 43-cm coelostat, a 56×32-cm elliptical flat mirror and a 30.5-cm objective doublet (made by Brashear) with a focal length of 18.3 m. Both mirrors were 30 cm thick and this helped reduce rapid distortion due to solar heating. The vertical path of the beam lessened the chance of thermal currents across the wave front. A laboratory contained a 9-m underground spectroheliograph which produced a 15-cm monochromatic solar image. When this instrument went into operation in 1907, it was found to be far superior to the Snow telescope.

With the vertical design now proved, Hale proposed a 150-ft (46-m) tower telescope in order to achieve an even larger solar image. Again the Carnegie Institution provided the funds, and construction began in 1909. This instrument was similar in design to the 60-ft telescope, except that the tower was of double construction: an outer tower protected the inner tower from the wind. The mirrors were fitted with water jackets for cooling, and the 30.5-cm objective was a triplet, designed to minimize the secondary spectrum. A 75-ft (23-m) spectroheliograph produced a 43-cm solar image. This instrument was finished in 1912, and after some alterations it lived up to expectations. All three solar telescopes on Mt Wilson, improved over time, have remained in use. Their value was quickly proved by the spectacular discoveries made by Hale and his staff.

In the meantime Hale had also acted on his plans for a large reflector. Shortly after the founding of the Observatory, the Carnegie Institution had approved the construction of a large reflector for spectroscopic studies of stars and nebulae. The 60-inch (152-cm) glass disc was transported from Yerkes to Mt Wilson's optical shops in Pasadena, where Ritchey began figuring it. The grinding and polishing machine was designed so that the turntable with the blank on it could be turned to the vertical, thus facilitating testing for zonal errors. Except for a four-month setback when scratches were discovered on the nearly finished mirror, Ritchey's work progressed smoothly, and the mirror was ready for silvering in August 1907.

The tube and mounting were designed by the Observatory's staff and built by the Union Iron Works in San Francisco. The 4.6-m tube consisted of eight steel tubes, rigidity being assured by a series of rings and cross-braces. The mirror was supported by a system of levers in a steel housing attached to the bottom of the tube. It was fork-mounted on the polar axle. Just below the fork a 3-m-diameter float resting in a mercury bath reduced friction. The telescope could be slewed by means of electric motors. The dome was of a double-wall design in order to minimize temperature variations to which the mirrors were so sensitive.

As in the case of the solar telescope, Hale designed the optical system of the 60-inch reflector so that the instrument could be used for a variety of purposes. As a Newtonian it was an f/5 instrument for photography and low-dispersion spectroscopy. In a modified Cassegrain configuration, using a convex hyperboloidal mirror at the prime focus and a plane mirror at the lower end of the tube to reflect the light to the side of the tube (an arrangement first used by Common in his 60-inch reflector), it could be used as an f/16 for spec-

8.3. The Mt Wilson 60-ft solar telescope (left), completed in 1907, and the 150-foot (right, with dome), completed in 1912. In 1914 a dome and vertical tube were added to the 60-ft. The building for the earlier Snow horizontal solar telescope is visible in the distance at the left.

trography and an f/20 for photography. Finally, as an f/30 coudé, light was reflected by an appropriately geared mirror through the hollow polar axis into a constant-temperature room housing a large spectrograph. This flexible optical system, which allowed the instrument to be used for visual, photographic and spectrographic purposes, was a model for future large reflectors.

The new instrument saw first light in December 1908. Although definition and rigidity were excellent, mirror distortion due to day–night temperature variations was a problem. Ritchey therefore equipped the telescope with a canopy of blankets to cover it during the daytime and covered the dome with a canvas screen (replaced in 1913 by a sheet-metal covering). The mirror support, too, required some improvements. By the summer of 1909 the reflector had been improved to the point at which

stellar images were round even at large zenith angles, and excellent direct photographs could be made even with exposures of several hours. The new instrument thus allowed great advances in direct photography and in the recording of stellar spectra, and by 1909 there could be little doubt that in astrophysics the reflector had replaced the refractor as the primary observatory instrument.

Yet, in size and usefulness this instrument remained unique for nearly a decade. Only the rebuilt Crossley reflector at Lick with its 36-inch (91-cm) aperture was comparable. At Harvard, E. C. Pickering was having no success in his efforts to turn Common's 60-inch reflector into a useful observatory instrument (see Chapter 3). Elsewhere a number of smaller reflectors were built in the first decade of the twentieth century, but with mixed success. In England, J. H. Reynolds provided a 30-

8.4. George Ellery Hale and Andrew Carnegie in front of the new 60-inch reflector, 17 March 1910.

inch (76-cm) Common mirror with a sturdy mounting and shipped the resulting instrument to Egypt for installation in the Khedival Observatory at Helwan, but this instrument was never truly satisfactory. The John A. Brashear Company in Pittsburgh and the Zeiss Foundation in Jena, however, were successful in supplying reflectors with apertures of up to 100 cm. Brashear made the

optical parts for 95-cm Cassegrains for the Lick observing station in Chile in 1904 and for the Ann Arbor (Michigan) Observatory in 1911. Zeiss supplied a 72-cm reflector (patterned on Ritchey's 24-inch (61-cm) instrument) to the Heidelberg Observatory in 1906, and a 100-cm reflector to the new Bergedorf Observatory near Hamburg in 1910.

8.5. An accident on the trail: transporting the central section of the 100-inch telescope tube up to Mt Wilson, *c.* 1916.

The 100-inch Hooker Telescope

While the 60-inch reflector was still under construction, Hale was already studying the feasibility of a much larger reflector, one with an aperture of 100 inches (254 cm). He was fairly confident that such an instrument could be successfully mounted: the question was whether a glass disc of this size could be made and whether funds for such an instrument could be obtained. In 1906 Hale induced J.D. Hooker, a wealthy Los Angeles merchant, to donate $45000 to the project, and the St Gobain glass works agreed to attempt the casting of a 100-inch disc. After several failures, an apparently satisfactory disc was cast and shipped to Pasadena. But when it arrived there in December 1908 Ritchey found it to be full of small bubbles, and he and Hale rejected it. Although St Gobain agreed to continue its efforts, the future of the telescope was now in doubt. When further

testing showed that there was a good chance that the disc could be figured without cutting into the subsurface bubble layers, Hale prevailed on the stubborn Ritchey to begin work on the mirror on a trial basis. With work now started, in 1910 Hale and Hooker managed to interest Andrew Carnegie in the project, and under Carnegie's inducement the Carnegie Institution agreed the next year to fund the instrument. Hale estimated its cost at $500000. When, by the end of 1911, Ritchey had finished the fine-grinding of the spherical curvature without encountering bubbles, the future of the 100-inch telescope seemed assured. The preliminary designs of the dome, drawn up by Pease and other members of the observatory's staff, were now turned over to the engineering firm of D.H. Burnham in Chicago for completion. The English yoke mounting was designed initially by Ritchey and then modified by Hale and Pease.

The polishing and testing of the mirror presented Ritchey with many problems. In order to eliminate dust in the shop (especially built for this project), water-spray air filters had to be installed. The polishing operation itself necessitated the construction of a 100-inch-diameter polishing tool. For testing, a 60-inch flat mirror had to be ground and polished, and the testing itself could not begin until a ventilating and heating system with thermostatic controls had been installed. Even then, the mirror could not be tested during the winter months. Initial tests of the spherical surface, in 1912, showed astigmatism, and it was feared that this was due to internal stresses in the disc. In view of the continued failures at St Gobain to produce a better disc, the future of the project was again cast in doubt. It was with hesitation that Hale approved the contract with the Fore River Shipbuilding Company of Quincy, Massachusetts, for the construction of mounting and dome. Not until the next year did Adams manage to prove that the astigmatism was caused by improper support of the mirror in the vertical position, and to rectify the defect.

Work now progressed steadily. By the end of 1914 Ritchey had achieved near perfect spherical curvature of the mirror. At this point W. L. Kinney took over the work. It took another two years of painstaking labour to parabolize the mirror. Near the end of 1916 tests showed that the mirror was a success. Its focal length was 12.88 m and it weighed 4082 kg. Silvering was done quickly and successfully. The finishing of the secondary mirrors was, however, delayed somewhat when, upon the entry of the United States in the war in 1917, military work took precedence at the optical shop.

In 1915 the Fore River Shipbuilding Company had begun sending components of the mounting to Pasadena. The heaviest parts were sent by ship; their transport to the top of Mt Wilson necessitated the building of a new road. Since the mounting had not been assembled prior to shipping, a number of fitting problems had to be solved during installation. Not until July 1917 was the mounting ready to receive the great mirror. In November the telescope was tested for the first time. The first look was discouraging, but later that night, when the mirror had cooled off, it was apparent that the instrument was a success. By 1919, after some minor problems had been eliminated, the Hooker telescope was in regular use, but the construction

of ancillary instruments has continued to this day. One of the most important, the coudé grating spectograph, did not come into full use until 1940. The Hooker telescope cost the Carnegie Institution $600 000, not including the labour of the Mt Wilson staff or the contribution of Hooker.

Optically, the Hooker telescope is very similar to the 60-inch reflector. It can be used as an f/5 Newtonian, and f/16 modified Cassegrain, or an f/30 coudé. In the latter case the beam is brought through the hollow polar axis into a constant-temperature room in the south pier for high-dispersion spectrography. The 100-tonne telescope is supported on the piers by means of mercury floatation bearings. The polar axis of the English yoke mounting (chosen, despite the fact that it restricts access to polar regions, in view of the instrument's great weight) is a rectangular framework of about 10×5 m. The 30-m double-skin dome is similar to that of the 60-inch reflector, and both dome and telescope are controlled by electric motors. In sum, except for the mounting and minor details, the 100-inch telescope is a larger version of the 60-inch instrument: but it represented a gain in light-gathering power of 2.8, and it enabled astronomers to establish the nature and distance of spiral nebulae and to measure their velocities. It therefore played a central role in the revolution in cosmology that took place in the 1920s.

The importance of the 100-inch telescope as a pioneer for a new generation of instruments is masked by the events of World War I and the ensuing economic difficulties in Europe. One other notable large reflector should, however, be mentioned. After having cast the less than perfect 100-inch disc, the St Gobain glass works had little trouble casting a $73\frac{1}{2}$-inch (187-cm) disc for the proposed Dominion Astrophysical Observatory near Victoria, British Columbia, Canada. This disc was shipped in July 1914, a few days before the outbreak of war, and a few weeks before the total destruction of the venerable plate-glass works. Brashear ground and polished the Cassegrain mirror, and Warner & Swasey of Cleveland built the English equatorial mounting and dome for the 72-inch (183-cm) telescope. A model of planning and execution, the new observatory went into operation in April 1918. The instrument performed excellently, and it allowed J. S. Plaskett and his

8.6. 100-inch reflector under construction, Mt Wilson, *c.* 1917.

small staff to make important contributions to astrophysics, notably in the study of interstellar absorption of light.

The 200-inch telescope

During the planning and construction of the Hooker telescope, Hale could not be entirely certain that the gain of the new instrument over the 60-inch reflector would in fact be in proportion to the increase in aperture. When the 100-inch telescope went into operation, however, his doubts were dispelled: the gain was not limited by atmospheric tremors. And when the instrument was equipped with a Michelson stellar interferometer with an effective aperture of 6 m, Pease was able to determine that with regard to atmospheric turbulence, reflectors with apertures up to 6 m would be entirely justified. Hale and his staff now began to

8.7. The Victoria 72-inch reflector, completed in 1918. J.S. Plaskett, director of the Dominion Observatory, stands beside the controls.

consider a telescope with an aperture of 200 or 300 inches (508 or 762 cm). After Hale had resigned the active directorship of the Mt Wilson Observatory in 1923 for reasons of health, he slowly began setting the wheels in motion for this project.

A tentative design by Pease showed that mounting such a large instrument was possible within the framework of existing mechanical technology. But two major problems needed to be solved: could a satisfactory disc of 200 or 300 inches be made, and could such a huge telescope be financed? The failure of the St Gobain glass works to produce a perfect 100-inch disc, as well as the sensitivity of ordinary glass to temperature changes, showed that a much larger disc would have to be made of a different material. The Hooker telescope had cost about six times as much as the 60-inch Mt Wilson reflector, indicating that for such pioneering instruments the cost increased as at least the third power of the aperture. A 200-inch telescope would cost more than $5 000 000! It would be difficult to find funds for such an instrument, especially since the technical feasibility remained in doubt.

Two glasses held out hope: pure silica and borosilicate glass. Pure silica (fused quartz) has a thermal expansion coefficient an order of magnitude lower than that of ordinary glass. But its high melting point and viscosity made it extremely difficult to cast pure silica in large shapes without trapping bubbles. Elihu Thomson at the General Electric Company had managed to cast discs of up to 30 cm, and there was some hope of making them larger. Borosilicate glass, first made by Michael Faraday, had been mass-produced since 1915 by the Corning Glass Company under the trade name 'pyrex'. Its thermal expansion coefficient was about a third as great as that of ordinary glass. Experiments were under way at Corning to make pyrex with a higher silica content and, therefore, a lower expansion coefficient. Indications were that pyrex could be cast into large shapes, but even with its relatively low coefficient of expansion a 200 inch disc of pyrex would take about nine years to cool after casting. A pyrex disc would have to have a ribbed structure with sections no thicker than about 10 cm if a reasonable cooling time was to be achieved.

Encouraged by these developments, Hale began his campaign for a larger telescope in 1928. His article entitled "The possibilities of large tele-scopes", written for *Harper's Magazine*, became the occasion for a series of discussions with W. Rose, president of both the General Education Board and the International Education Board (organizations associated with the Rockefeller Foundation). Rose, an enthusiastic promoter of 'big science', had been aware of Hale's dream for some time. While Hale asked for funds for a feasibility study only, Rose wished to make a grant for the entire telescope. After consulting with engineers, Hale decided that a 200-inch aperture was a more realistic target than the 300-inch aperture advanced by Pease. After some difficulties over the question as to which institution was to be the recipient of the instrument, the International Education Board awarded $6 000 000 to the California Institute of Technology (Caltech) in Pasadena. There was, however, one important condition: the grant was for the construction of the telescope and ancillary equipment only, and before construction could begin Caltech had to secure an endowment adequate for meeting operation costs. When Henry M. Robinson, a Los Angeles banker, agreed to provide an appropriate endowment, the International Education Board gave its final approval for the grant in the autumn of 1928.

The design and construction of the giant telescope promised to be a difficult and very complex task. New designs and technical innovations were required in order to combine a number of new (and in some cases not yet existing) technologies in a new form. No single man or company could coordinate and execute a project that drew on technical areas as diverse as glass-making, mechanical and electrical engineering, optics, seismology and meteorology. This was 'big science'; it required a committee approach. A four-member Observatory Committee, headed by Hale, was put in overall charge. It was aided by a nine-member Advisory Committee made up of astronomers and other scientists from a number of institutions. There were committees on the mirror, the site and the mounting. Other experts were appointed as directors and executive officers. In this manner Hale managed to enrol the best scientists and engineers in the United States in the effort. J. A. Anderson (1876–1959) was appointed executive officer in charge of construction as well as optical work.

The first problem was that of the mirror. On its

solution depended the ultimate success, indeed, the very possibility, of the telescope. After evaluating a number of suggestions, Hale and the Observatory Committee asked the General Electric Company to attempt to make a fused-quartz disc; the estimated cost, in 1928, was $250000. Thomson and his staff planned a series of discs of increasing size, culminating in a 200-inch disc. But in 1928 only small quartz discs had been made, and the leap to 200 inches posed a great problem. When it proved to be impossible to eliminate bubbles from the molten silica, clear quartz plates were welded onto the surface of the discs, and when this proved to be impossible for larger discs, Thomson tried spraying the disc surfaces with quartz dust liquified by means of an oxy-hydrogen flame. With great difficulties General Electric managed to make a good 22-inch (56-cm) disc in 1929. By July 1931 two 60-inch discs had been made, but they were of questionable quality, and by this time $639000 had been spent on the fused-quartz experiments.

In the meantime, Pease had discovered that a large part of the flexing of the 100-inch mirror of the Hooker Telescope was due to inadequacies of the mirror support, not to thermal expansion and contraction. When a new ball-bearing support system was installed in 1931, the mirror performed much better, and it now appeared that pyrex would be a perfectly suitable material for the 200-inch mirror. Hale therefore terminated the experiments at General Electric and, in October 1931, contracted with Corning for a pyrex disc.

The largest pyrex disc cast by Corning up to this time was 30 inches (76 cm) in diameter. It was planned to make discs of 30, 60, 120 and then 200 inches; the 120-inch disc would be used for testing the 200-inch mirror, and the other discs were to be used in the telescope itself. All discs were to have a deeply ribbed structure to provide for a sufficiently short thermal time constant and to minimize the weight of the larger mirrors. This meant that the moulds had to be provided with cores, and this proved to be a problem: cores broke loose when their anchoring bolts were weakened by the high temperatures. In spite of these difficulties, work at Corning progressed smoothly. A 26-inch (66-cm) disc was followed by a 30-inch disc for the coudé flat mirror. Then came the 60-inch Cassegrain blank and, in June 1933, the 120-inch testing disc. The first attempt to cast the 200-inch disc was

made in March 1932, but several cores broke loose. Finally, on 2 December 1934, a perfect 200-inch disc was cast. Cooling and annealing took a further year; transportation to Pasadena required a specially built railway car. Tests of the disc showed that internal stresses were extremely small. The ribbed structure kept the weight below 20 tonnes.

Even before the 200-inch disc was cast, the new glass technology began to have an impact on telescope building, and nowhere is the importance of the 200-inch telescope as a trail-blazing instrument better illustrated. Since the destruction of the St Gobain glass works in 1914, only Zeiss in Jena and the Parsons Optical Glass Company in Derby, England, had been able to supply large mirror blanks. Zeiss's 125-cm reflector for the Berlin-Babelsberg Observatory, mounted in 1927, and the Grubb-Parsons 40-inch (102-cm) reflectors for the Simeis Observatory in the Crimea (1928) and the Saltsjobaden Observatory near Stockholm (1931) were their largest respective efforts. The only disc larger than 50 inches (127 cm) cast during this period was the 69-inch (175-cm) borosilicate-glass disc cast as an experiment by the US Bureau of Standards in 1928. It was figured by J. W. Fecker (the successor of Brashear) and mounted in 1931 by Warner & Swasey for the Perkins Observatory of Ohio Weslyan University. For the 61-inch (155-cm) reflector for Harvard's Oak Ridge Station in Massachusetts, Fecker could do no better than refigure one of Common's 5-ft mirrors. The difficulty of obtaining discs larger than about 125 cm was, therefore, the crucial technical obstacle to the proliferation of very large telescopes.

When a 74-inch (188-cm) reflector was ordered in 1930 from Grubb-Parsons for the David Dunlap Observatory near Toronto, the Parsons Optical Glass Company failed in its attempts to cast the disc. By 1932 Corning had succeeded in casting the 60-inch disc for the 200-inch telescope, and the Company now agreed to cast a 74-inch disc for Grubb-Parsons. It was cast a few days before the 120-inch disc, in June 1933, and the Dunlap reflector went into operation in 1935. In 1938 Corning cast a 76-inch (193-cm) disc for Grubb-Parsons for a 74-inch reflector to be erected near Pretoria, South Africa, the southern station of Oxford's Radcliffe Observatory. Delayed by the war, this instrument went into operation in 1948.

Corning also supplied discs for other large

8.8. Transport of the 200-inch mirror, Pasadena, April 1936.

reflectors. In 1937 Fecker replaced the 61-inch mirror of the Oak Ridge reflector with a pyrex mirror, and he also figured a pyrex mirror for the 60-inch reflector built by Warner & Swasey for Argentina's Córdoba Observatory (1940). In 1933 Corning cast the disc for the 82-inch (208-cm) telescope for the new McDonald Observatory in Texas. This instrument, built by Warner & Swasey, went into operation in 1939. A 98-inch (250-cm) disc supplied by Corning to the University of Michigan remained idle until it became the primary receptor of the Isaac Newton Telescope at the Royal Greenwich Observatory at Herstmonceux in 1968. The testing disc for the 200-inch mirror became the mirror for the 120-inch (305-cm) reflector of the Lick Observatory in 1959. Clearly, solving the problem of the mirror for the 200-inch telescope was an important turning point for very large telescopes in general.

During this same period reflectors also benefited from another innovation. Around 1930 it became

possible to deposit thin layers of metals on surfaces in a high vacuum, and investigations showed that aluminium was far superior to silver as a coating for telescope mirrors: it has a much higher reflectivity in the ultraviolet region. Once covered by a very hard, air-tight, and completely transparent layer of oxide, the aluminium coating lasts for five years or more, and it can be washed with soap and water. Silvered mirrors tarnish so rapidly that they have to be burnished frequently, leading to light-scattering, and they have to be resilvered about twice a year. Making high-vacuum installations large enough to accommodate telescope mirrors, and achieving satisfactory adhesion of the aluminium coating, presented technical problems that had to be solved. By 1933 John Strong of Caltech had managed to aluminize the secondary mirrors of the 60-inch and 100-inch Mt Wilson reflectors. In December of that year he aluminized the 36-inch mirror of Lick's Crossley reflector at Mt Wilson, and by the end of 1935 he had done the

8.9. Mirror of the 200-inch telescope in the Pasadena optical shop.

same for the primary mirrors of the 60-inch and 100-inch telescopes of Mt Wilson. Gradually, over the next two decades existing mirrors all over the world were aluminized.

In the meantime, work on the 200-inch telescope had progressed steadily. The choice of a site had been approached with great care. A number of observers had made measurements of the weather conditions and seeing at many possible sites in the southwestern United States. In 1934 Palomar Mountain in southern California was chosen: it had the virtue of being close enough to Pasadena to allow fairly convenient access, yet far enough away from the city lights and pollution for delicate work on nebulae. In September 1934 land was purchased on the mountain and the San Diego county authorities began building a suitable road to the top.

As long as the question of the feasibility of the main mirror was not answered, the mounting remained in the design stage. In 1928 E. P. Burrell, the chief engineer of Warner & Swasey, who had the 72-inch Victoria and the 69-inch Perkins reflectors to his credit, came to Pasadena to work with the staff of Mt Wilson and Caltech. The choice was between an open-fork equatorial, and a yoke-type equatorial. Initially the fork design was preferred: it allowed access to the pole, and Burrell and Pease felt that the rigidity problems could be overcome. Tests of a scale model were promising. In 1932, however, a new yoke design was submitted in which the problem of polar access was solved. If the north bearing of the polar axis was made in the shape of a very large horseshoe, the telescope could be lowered into it far enough to point to the Pole. After a model of this design had been tested, it was approved, and in 1936 the Westinghouse Company in Philadelphia was awarded the contract for the construction of the mounting. Several large components would, however, have to be subcontracted.

For reasons of field size, optical speed and scale, Hale and Pease from the beginning planned a mirror with a very low focal ratio, f/3.3 as it turned out. An added advantage of such a configuration was that it meant a shorter tube, hence a smaller dome. At the primary focus where much of the photographic work is done, however, this low focal ratio meant that coma would be a much greater problem than it was in the 60-inch and

100-inch telescopes. The diameter of the field of good definition at the primary focus would be about a centimetre. Hale asked F.E. Ross (1874–1960) of the Yerkes Observatory to design a corrector lens to eliminate coma over a larger field. The Ross corrector lenses made for the 60-inch and 100-inch reflectors were very successful and Ross now designed several correctors for the 200-inch instrument. Ross lenses have become standard equipment for large reflectors.

The 200-inch telescope was the first telescope large enough to allow an observer's cage in the tube at the primary focus without cutting off more light than the Newtonian mirror. The design called for a 72-inch-diameter cage carrying the secondary mirrors and the Ross lens to be centred in the top of the telescope tube. The weight of the mirror at the bottom and the cage at the top presented the designers with a problem: in order to function effectively the Ross lens must be collimated with the primary mirror within very narrow tolerances, and calculations showed that at various positions the tube would flex more than these limits. The problem was solved by Mark Serrurier, a Caltech graduate who had worked on the Golden Gate Bridge. Serrurier designed parallelogram trusses, connecting the top and bottom rings of the tube to the square section housing the bearings of the declination axle. When the mirror cell and observer's cage were attached to the end rings, flexure was such that the optical axes of mirror and corrector lens had only translatory motion; their rotating motion with respect to the axis of the tube was well within the allowable limits. This design, called the Serrurier truss, is now standard in large telescopes.

The giant horseshoe bearing on the north pier, and its smaller counterpart on the south pier, were to support a total weight of about 500 tonnes. Yet they had to allow smooth and precise motion with almost no friction in order to prevent the polar axis yoke from twisting under large forces. Ball-bearings would not reduce friction sufficiently, and mercury-floatation bearings were ruled out in view of the great weight.

Anderson found the solution in a new development in the machine-tool industry: oil pad bearings. The polar axis is floated on a thin film of oil. The pads consist of Babbitt metal. Oil under a pressure of 300 psi (0.2 kg/mm^2) is forced through

8.10. Cutaway sectional view of the 200-inch telescope and its dome, part of a famous series by Russell W. Porter, 1938. An observer is shown at the prime focus cage near the top of the tube, and the coudé room is to the left below the south bearing of the mounting.

orifices in the pads, and the journals ride across them on an oil film 0.076 mm thick. The horseshoe has an outside diameter of 14 m, and it was made in three sections weighing about 50 tonnes each. Its flexure under load was calculated in advance, and before machining its shape was distorted by an appropriate amount, so that when the telescope was far east or west the horseshoe's outside surface would be exactly circular.

In view of the flexibility of the main mirror, the mirror support system had to be elaborate and extremely accurate. The mirror was carried on thirty-six support mechanisms inserted in the holes in the mirror. Elaborate lever arm mechanisms balanced the changing forces on the mirror as its position varied. When the mirror arrived in Pasadena, the holes for the support system were ground to the proper size and the mirror was mounted on the support system during figuring and testing.

When Hale died in the summer of 1938, the figuring of the main mirror was progressing steadily at the optical shop in Pasadena under the direction of Marcus Brown. The 137-ft (42-m) dome, designed by R. W. Porter, had already been erected on Palomar Mountain. It had a welded outer skin of $\frac{3}{8}$-inch (1-cm) steel plate, a 4-ft (122-cm) ventilated air space and a 4-inch (10-cm) inner skin of steel and aluminium boxes filled with crumpled aluminium foil. This arrangement kept temperature variations at the main mirror to a degree or two.

Although some had expressed the hope that the telescope might be finished by 1940, this was not to be. Figuring the main mirror, testing it at various stages, and finding the sources of errors (for example, in the support system), took longer than had been anticipated. World War II brought work to a standstill. The final parabolizing under the direction of I. S. Bowen did not start until 1945, and not until November 1947 was the almost finished mirror transported to the top of Palomar Mountain where the last adjustments were made.

Optically the 200-inch telescope is similar to the 100-inch and 60-inch instruments. The focal ratio

of the main mirror is f/3.3. At the prime focus with a Ross corrector in place the ratio is slightly larger. The two Ross lenses most frequently used give ratios of f/3.6 and f/4.7. With one hyperboloidal mirror in the Cassegrain configuration the image is formed about 5 ft (1.5 m) behind the front surface of the main mirror, giving a ratio of f/16. With this mirror the rays can also be reflected by a plane mirror to form an image in the eastern arm of the yoke. Two other hyperboloidal mirrors can each be moved into position to form an image at the coudé focus on the polar axis below the south bearing. The construction of ancillary equipment continues to this day.

Postscript

Named the Hale telescope at its dedication on 3 June 1948, the 200-inch reflector was the prototype of large post-war telescopes, and the third quarter of the century saw the steady proliferation of instruments with apertures of over 2 m, the largest being the 6-m Soviet telescope completed in 1975. All incorporated features that had been first developed in the design of the Hale telescope, such as low expansion glass. Indeed, Hale's enduring accomplishment lies almost as much in the lead he gave to others in the solution of technical and financial problems in the building and managing of great telescopes as in the scientific results achieved at his observatories.

Further reading

I. S. Bowen, The 200-inch Hale telescope, in Gerard P. Kuiper and Barbara M. Middlehurst (eds), *Telescopes* (Chicago, 1960; vol. 1 of the series *Stars and Stellar Systems*), 1–15

Henry C. King, *The History of the Telescope* (London and Cambridge, Mass., 1955)

Helen Wright, *Explorer of the Universe: A Biography of George Ellery Hale* (New York, 1966)

Helen Wright, *Palomar, The World's Largest Telescope* (New York, 1952)

Helen Wright, Joan N. Warnow and Charles Weiner (eds), *The Legacy of George Ellery Hale* (Cambridge, Mass., 1972)

Additional information can be found in the *Carnegie Institution of Washington Yearbook*, an annual publication beginning in 1902.

9

Astronomical institutions in the southern hemisphere, 1850–1950

DAVID S. EVANS

Introduction

The development of observatories in the southern hemisphere has been determined as much by geographical and political factors as by scientific ones. The southern terrestrial hemisphere is largely oceanic: the parallel at latitude 45°S includes only about 1000 km of land, and even at latitude 30°S only 22% lies on land. South Africa, Australia and South America, like the isolated islands that also have some astronomical history such as St Helena, Mauritius and Indonesia, are all ex-colonial territories and astronomy in them owes nothing to aboriginal indigenous culture. On the contrary, cultural links have almost invariably been back to the colonial power or its successor. Although in general the southern sites are climatically superior to the older northern ones, southern astronomers have been few: in 1950 only one in twenty of the members of the International Astronomical Union had southern addresses, and only in the late twentieth century are there signs of the emergence of a specifically southern-hemisphere astronomical initiative with mutual links facilitated by new air routes joining the major centres.

For these reasons development of southern-hemisphere astronomy has been much slower than in the north, although, ironically, the southern sky includes some of the most important celestial features – the galactic centre, the majority (and brightest) of the globular clusters, and the Magellanic Clouds, to cite only the most outstanding.

The sequence of events in astronomy in the south has been similar in all areas although performed at a different pace and to a different completeness in the different territories. It always begins with the demands of navigation and the need for the provision of a time service for navigators, and with the search for improvement in the determination of geographical positions. If the means are there, the next stage is a survey of southern star positions with, in some cases, applications to geodetic and cadastral survey. In a few instances we see a transition to purely astronomical data – star positions and catalogues, proper motions, parallaxes; later, the beginnings of astrophysical data, such as photometry, the discovery of variable stars and the determination of radial velocities.

In most cases there was an element of scientific colonialism. The southern observers, however well qualified for true research, were overworked and under orders to provide the data for analysis back home. Leisure and the glory of analysing the data were the prerogative of home-based astronomers, and the southern observers, being regarded as mere data gatherers, were even assigned inferior instruments though their observational opportunities were by far superior. There is, in the century under review, an undercurrent of resentment that southern observers were never treated as equals by those in the north by whom their efforts were directed.

Only in relatively modern times has true scientific initiative in astronomy developed in the south. Two examples come in the middle of our period: the discovery of variables in great numbers by S.I. Bailey, and the far-reaching initiatives of David Gill, who influenced astronomy for the next four-score years in spite of the opposition at home in Britain.

Southern-hemisphere astronomy approaches maturity only at the very end of our period, and to trace how this came about we must recount the story in terms of the diverse developments region by region, starting with South Africa, which has had the longest astronomical history.

South Africa

The Royal Observatory at the Cape of Good Hope
In South Africa astronomy already had a history when our period begins. Halley (from 1676 to 1678 at St Helena) and Lacaille (from 1751 to 1753 in South Africa) had surveyed the southern sky, the latter producing a star catalogue of respectable length and accuracy. The British had established the Royal Observatory at the Cape in 1820, intended to be a southern Greenwich but always inadequately financed and equipped. The director appointed in 1834 was Thomas Maclear (1794–1879), originally an amateur who had trained as a surgeon. In the same year John Herschel had begun his intensive four-year stay at the Cape and with his surveys of double stars, clusters and nebulae had advanced knowledge of the southern sky to a degree comparable to that in the north. After his return to Britain in 1838, Herschel was to lobby throughout his life to promote the interests of the Cape Observatory.

Maclear's instructions from the Admiralty would have taxed the staff of a far larger establishment than the team of four assistants he finally secured, but his instrumental resources were enhanced in 1854 by the arrival from England of a new transit circle by Troughton & Simms. During Maclear's time large numbers of astrometric observations were made but reductions fell into arrears despite the urging of G. B. Airy at Greenwich. Maclear did complete a monumental investigation of Lacaille's anomalous measurement of a degree of the meridian; he examined the relevant historical documents; and he repeated and greatly extended Lacaille's geodetic work insofar as that was possible. The two-volume *Verification and Extension of Lacaille's Arc of Meridian* published in 1866 marked the beginning of serious South African geodesy.

Persuaded to retire in 1870, Maclear was succeeded by a professional, E. J. Stone (1831–97), who had what was to become in the twentieth century a standard set of qualifications – the Cambridge Mathematical Tripos and a Chief Assistantship at Greenwich. Stone concentrated on reduction of the accumulated observations and after his resignation in 1879 to become Radcliffe Observer at Oxford brought the 12400-item star catalogue to completion. However, this policy left the observatory

itself in a run-down condition for his successor, David Gill (1843–1914).

Although Gill had been a student under J. Clerk Maxwell at Aberdeen, he came to astronomy as an amateur while engaged in the family watchmaking business. His training enabled him to become one of the outstanding observers of the nineteenth century. His favourite instrument was the heliometer, a refractor with the objective sawn in half across a diameter: relative displacement of the two halves by means of a screw mechanism produced double images and enabled the observer to measure accurately small angular displacements of stars with respect to their neighbours.

After leaving the family business, Gill had first served as astronomer at the private observatory of Lord Lindsay at Dun Echt in Scotland. While there, he first went south in 1874 to observe the transit of Venus from Mauritius and to determine the solar parallax. Dissatisfied with these results, he turned his attention to Mars and the minor planets, and in 1877 spent six months on Ascension Island.

On his appointment to the Cape in 1879, Gill arrived to find the observatory disorganized and dilapidated. But by the time of his retirement in 1907, his activities had brought the Cape to full stature as an independent scientific institution capable of initiating and completing its own projects, in several cases financed out of Gill's own pocket or from donors quite unconnected with the Admiralty in London. In collaboration with a visitor, W.L. Elkin, who was later to become director at Yale, Gill used the heliometer to determine the first nine reliable annual parallaxes of stars in the south, and from observations of Iris, Victoria and Sappho he derived a value (8.802 ± 0.005 arc seconds) for the solar parallax superior in accuracy to the efforts of his successors almost until the introduction of radar methods.

Gill's greatest influence on the whole world of contemporary astronomy began in 1882 and, for better or worse, was to continue for eighty years. Dry plates had then just come into use and Gill, with the aid of a local photographer, obtained an excellent photograph of the Great Comet of 1882. This also showed large numbers of star images and Gill immediately conceived the possibility of making star charts by photography. This led first to the production of the *Cape Photographic Durchmusterung (CPD)*, a roster (on the lines of the *Bonner*

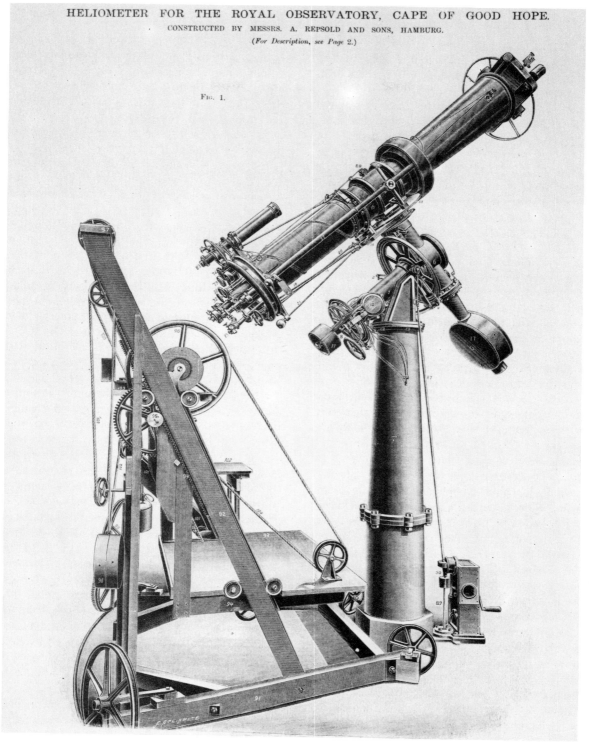

HELIOMETER FOR THE ROYAL OBSERVATORY, CAPE OF GOOD HOPE.

CONSTRUCTED BY MESSRS. A. REPSOLD AND SONS, HAMBURG.

(For Description, see Page 2.)

Fig. 1.

9.1. Gill's 7-inch heliometer, 1887, probably the best and certainly the most widely-used heliometer ever made.

9,2. The Great September Comet of 1882 photographed at the Royal Observatory, Cape of Good Hope. This 100-minute exposure on 7 November 1882 shows an 18-degree tail and stars to the tenth magnitude. This plate provided the impetus for photographic stellar mapping.

Durchmusterung (*BD*)) of all stars down to a given magnitude limit south of declination $-18°$. The work became possible only through the cooperation of the Dutch astronomer J.C. Kapteyn (1851–1922) with whom Gill devised the measuring machine. All the plates were measured in the Netherlands, with assistance from local convicts, and Gill personally financed much of it since W.H.M. Christie, the Astronomer Royal, though originally supportive, blocked funds for its completion. The *CPD* was published in three large volumes (1896–1900) enumerating 454 875 stars with estimated magnitudes and positions to the same approximations as the *BD*.

Gill's early photographs had excited the interest of the Henry brothers in Paris and with the collaboration of the director of the Paris Observatory, an international Astrographic Congress was held there in 1887 (see Chapter 2). The sky was parcelled out into zones to be observed by various observatories around the world. The subsequent history of the project, which dragged on until the 1960s, makes one wonder whether it should be regarded as Gill's greatest triumph or his grandest failure. It was certainly the peak of his international influence, but while some observatories finished their zones promptly, elsewhere there are endless stories of zones transferred to more efficient institutions either for observation and reduction or for reduction only. The work in general was overtaken by improved methods before it could be completed and little use is now made of the astrographic catalogue volumes themselves.

The long list of visitors attracted to the Cape by Gill's reputation includes J. Franklin-Adams (1843–1912), who made photographic charts of the southern sky and donated his cameras to the Union Observatory at Johannesburg, and R.T.A. Innes (1861–1933). Innes, a British-born wine merchant from Australia, joined the Cape as a clerical assistant in 1896, but spent his time revising the *CPD*, being an unparallelled double-star observer and a connoisseur of recondite mathematics. He was to become first director of the Union Observatory.

At the end of Gill's time F. McClean endowed the Victoria telescope, a 24-inch (61-cm) photographic refractor with an 18-inch (46-cm) photovisual on the same mounting. The telescope was used initially to measure radial velocities to check the solar parallax. Gill was succeeded in 1906 by his chief assistant S.S. Hough (1870–1923), who had as his chief assistant J.K.E. Halm (1866–1944), a naturalized German from Edinburgh. Both were astronomers of high theoretical attainments who showed their mettle with a series of papers on the distribution of stars in the Cape Zone and on star streaming, which had been announced by Kapteyn at the South African meeting of the British Association for the Advancement of Science in 1905.

Hough was succeeded in 1924 by H. Spencer Jones (1890–1960), who revived the somewhat disorganized observatory by recruiting able South Africans and instituting vigorous programmes until his departure on appointment as Astronomer

9.3. Gill's original arrangement for the *Cape Photographic Durchmusterung* telescope, as first mounted in November 1886. The 9-inch object glass by Grubb, presented by James Nasmyth, was corrected for photographic work. The 6-inch Dallmeyer lens below had a shorter focal length and was intended for photography of fainter stars. The 5-inch Dolland guide telescope came from a disused transit instrument.

Royal in 1933. The Cape Astrographic Zone was reobserved to give 40 000 good proper motions and the Victoria telescope was turned over to the determination of trigonometric parallaxes by photographic methods. Spencer Jones initiated the *Cape Photographic Catalogue* of the southern sky, containing precise positions and proper motions. The *Catalogue* was based on wide-angle plates with reference star positions controlled by transit observations at the same mean epoch. The work was to be a major preoccupation of the observatory until well into the second half of the twentieth century. Spencer Jones also organized the international Eros programme for the determination of the solar parallax, which will be discussed in Chapter 13.

John Jackson (1877–1958), Spencer Jones's successor and a former Greenwich Chief Assistant, concentrated almost entirely on projects initiated by Spencer Jones. At the end of his administration in 1950, 1600 parallaxes had been determined. An important feature of the *Cape Photographic Catalogue* was the determination of magnitudes and colours of the 70 000 stars it would contain, a task assigned to R. H. Stoy, who became Chief Assistant in 1935.

The Union Observatory

When in 1903 Innes became, on Gill's recommendation, the director of the meteorological observatory just outside Johannesburg, he took with him a small astronomical telescope. Two years later Gill, Kapteyn and J. O. Backlund persuaded the Transvaal Government to establish there an astronomical observatory, and this became the Union Observatory when the separate colonies were united in 1910. Innes's double-star observations improved still further with the arrival in 1925 of a 26-inch (66-cm) visual refractor, which was to be at the centre of double-star work in the southern hemisphere. The Franklin-Adams cameras were used to produce charts for the whole sky to beyond the sixteenth magnitude and also charts to match the *CPD*.

An extended stay by W. de Sitter at the Cape led to a permanent connection with Leiden so that at various times E. Hertzsprung, W. H. van den Bos, H. van Gent, A. J. Wesselink and others from Leiden spent time at Johannesburg. Van den Bos became director of the observatory (1941–56), and

his discoveries of 3000 new double stars and the observations of his predecessors are incorporated in the all-sky index catalogue later published by Lick Observatory.

The cooperation with Leiden led to the installation of Leiden's own 16-inch (41-cm) double astrograph funded by the Rockefeller Foundation and inaugurated by Hertzsprung in 1939. At the end of the period under review it was used especially by Theodore Walraven for precise sky-compensated observations of variable stars.

Durban

The union of the separate colonies in 1910 not only enhanced the status of the Union Observatory, but led to the demise of another, the Natal Observatory at Durban. It had been started by E. N. Nevill (1849–1940), a Fellow of the Royal Society with considerable standing as a celestial mechanician and lunar observer, who had the curious idea that an occupation such as astronomy should be conducted under the pseudonym of Edmund Neison. He arrived in 1882 with an 8-inch (20-cm) Grubb telescope, and five days later observed the transit of Venus. At one time he had four assistants, but financial support declined and the Observatory was closed in 1913.

Other institutions

F. Schlesinger, director at Yale, visited Johannesburg in 1923 and decided to install a photographic refractor of 36-ft (11-m) focal length for parallax work. By 1953, 1728 determinations of stellar parallax were made, at which time the station was moved to Canberra, Australia.

In 1924 W. J. Hussey, director at Ann Arbor, Michigan, visited South Africa to pick a site for what would become the Lamont-Hussey Observatory at Bloemfontein. In 1926 Hussey and R. A. Rossiter (1886–1977) departed for South Africa, but Hussey died of a heart attack in London. Rossiter ran the observatory from 1928 to 1952, with an individual record of the discovery of 5534 new pairs of double stars.

The decision to move the Boyden Station of Harvard University from Arequipa, Peru, to South Africa in 1927 resulted from the climatic reports by S. I. Bailey in 1908. J. S. Paraskevopoulos (1889–1951), a Greek immigrant to the USA who had been superintendent of the station at Arequipa since 1923, re-established it a few miles north of Bloemfontein. The equipment comprised a 60-inch (152-cm) Fecker reflector with a mirror by Common (which was too thin, presenting problems of effective mounting) and the Bruce, Boyden, Bache and Metcalf telescopes. Because of the depression, few astronomers and students came from the parent institution, but Paraskevopoulos and his few local helpers sent back to Harvard a vast number of plates, many useful for tracing the variability of stars and quasars. Second epoch plates were also taken for W. Luyten's monumental proper motion survey and during the 1940s the South African G.G. Cillié determined spectral types of several thousand stars for the *Cape Photographic Catalogue* using Boyden material.

Harold Knox Shaw brought about the transfer of the Radcliffe Observatory from Oxford to Pretoria, following his attendance at the South African meeting of the British Association for the Advancement of Science in 1929. Site testing at Pretoria was undertaken in 1930 by the British amateur, W.H. Steavenson, but progress was delayed by objections in Oxford to the removal of the Observatory to another country. A 74-inch (188-cm) reflector was ordered, with the intention that the initial programme would concentrate on radial velocities of southern B-stars, to provide all-sky data for these objects for use in studies of galactic rotation. The main mirror was to be cast at Corning, New York, and figured by Grubb-Parsons in England. But these plans went seriously awry. A successful disc was cast only just before the outbreak of war in Europe and it spent the war years buried under sandbags at Newcastle-upon-Tyne. Only in 1948 was it delivered to Pretoria, which then possessed the largest telescope in the south and became by far the most important southern observatory. Directed by A.D Thackeray (1910–77), the work began with the new scale of the universe based on observations of Cepheids and RR Lyrae stars in the Magellanic Clouds.

Indonesia

At a latitude of 6°S the Bosscha Observatory at Lembang is especially associated with J.G.E.G. Vôute (1879–1963). Born in Java, then a Dutch colony, he became a civil engineer before joining the Leiden Observatory in 1908 where his interest

in double stars took him to both the Cape and Union observatories. Returning to Java in 1919 he secured backing from wealthy friends for building, at an altitude of 1300 metres, the Bosscha Observatory which he equipped with a 60-cm Zeiss refractor. Vôute served as director until 1939. Subsequent directors included C.H. Hins and G.B. van Albada, both known for variable star work.

Australia

As elsewhere in the southern hemisphere, observatories in the separate colonies of New South Wales, South Australia, Victoria and Western Australia were initially remote from parent Britain and from each other. Each carried on navigational, meteorological and other observations until the establishment of the Commonwealth in 1900, after which many services were centralized. Most of the separate observatories were closed and replaced by a scientifically developed central organization, which, however, suffered vicissitudes during periods of economic depression and war.

New South Wales
Scientific astronomy began early under the influence of T.M. Brisbane, Governor and amateur astronomer at whose Parramatta Observatory (established in 1822) C. Rumker and J. Dunlop made notable surveys of southern star positions and southern double stars, which were published in 1835 as the *Brisbane Catalogue*. By 1848, however, Dunlop was in the last year of his life and the observatory in a sad state of disrepair. By 1850 Sydney had a university and the precursor of the Royal Society of New South Wales, and legislative authority was given for the establishment of an observatory on a site overlooking the harbour.

Following the standard pattern, W. Scott (1825–1917), a Cambridge mathematician recommended by Airy, arrived in 1856 and took over what remained of the instrumental armoury of Parramatta, and the usual meridian observations were undertaken. He was succeeded in 1862 by G.R. Smalley (1822–70), yet another Cambridge man, who came overloaded with diverse instructions from Airy. The post had been previously refused by a local amateur, J. Tebbutt (1834–1916), who was far more accomplished than many a professional and by 1908 could list 371 publications in various learned journals. With small instruments he observed comets and immense numbers of minor planet positions of such accuracy as to be prized by orbit computers.

Smalley was in turn succeeded by his assistant, H.C. Russell (1836–1907), a native New South Welshman. Russell, an early experimenter with astronomical photography, organized an extensive observing campaign for the 1874 Venus transit. In 1887 he attended the Astrographic Congress and accepted the enormous tasks of the 52° to 64° south declination zone for Sydney and 65° south to the Pole for Melbourne. Sydney Observatory, after many years, completed its assignments in 53 published volumes of results, but the task had proved so onerous that in 1898 a joint bureau for measuring the Melbourne and Sydney plates was set up in the former, supported by the two state governments. Russell's resignation on health grounds in 1905 was followed by an interval of uncertainty while the newly established Commonwealth Government discussed the status of meteorological and astronomical research.

In 1912 W. E. Cooke (1863–1947), who had been appointed first director of the observatory at Perth, Western Australia in 1896, became Government Astronomer and Professor of Astronomy at Sydney University, but his plans to move the observatory to a better astronomical site were frustrated by the effects of World War I. Cooke became expert in photographic astrometry, produced a catalogue of 1068 intermediate reference stars, and planned a mirror transit of advanced design that was never in fact constructed. Economic depression imposed restrictions and on his retirement in 1926 most of the observatory staff were transferred to other departments. The Melbourne Observatory was closed as a state institution in 1944 and Sydney took over its share of the astrographic catalogue.

Victoria

In 1852 R.L.J. Ellery (1827–1908), who had trained as a surgeon and emigrated to Victoria, urged on the Governor the need for an observatory for time service purposes. This advice was accepted and a small observatory was established at Williamstown near Melbourne, and Ellery himself became its director, a post he held for 42 years. The observatory was moved to South Yarra in 1863 and the Board of Visitors began an agitation for "the erection of a telescope of large optical power".

In 1852 a resolution of the British Association for the Advancement of Science led to the formation in London of a 'Southern Telescope Committee' to erect a telescope to be used for reobservation of John Herschel's southern clusters and nebulae. In 1865 the Government of Victoria, acting on the recommendation of a committee of the Royal Society of London, appropriated the money for what became known as the Great Melbourne Reflector, a long-focus instrument with a 48-inch (122-cm) speculum primary erected at Melbourne in 1869. Except for the crudely-mounted reflector at Birr Castle in Ireland, this was then the largest telescope in the world but (as described in Chapter 3) it proved a disappointment, with problems of operation, maintenance and administration restricting its effectiveness, though some drawings of nebulae were made. After some reconstruction the instrument ended up at the Commonwealth Observatory at Mt Stromlo.

Mt Stromlo Observatory

This, originally the Commonwealth Solar Observatory, owed its creation to the initiative of W. G. Duffield, an Australian who spent much of his life in England. He conceived the idea at an Oxford meeting of the International Solar Union in 1905 and secured the support of a strong group of British and Australian scientists. Offers were made of a 6-inch (15-cm) Grubb refractor by a trustee of Lord Farnham and of a 9-inch (23-cm) Grubb refractor from the private observatory of Colonel J. Oddie. After delays due to World War I, a British committee selected Duffield to erect and direct a solar observatory on a site already selected on Mt Stromlo near Canberra. The observatory acquired a 30-cm solar telescope fitted with a three-prism spectrograph, and in 1934 the solar astrophysicist C. W. Allen published an atlas of the solar spectrum from 4036 to 6600 Å. In 1930 an atmospherics research station was established on Mt Stromlo in conjunction with the Radio Research Board of the Council for Scientific and Industrial Research; A. R. Hogg was active in this department in cosmic ray and ionospheric research.

Richard v.d.R. Woolley, later Astronomer Royal, became director in 1939; he proposed to take advantage of the observatory's location for astrophysical work. In 1924 the well-known British amateur, J. H. Reynolds, had donated one of his

9.4. The Great Melbourne Reflector with its 48-inch speculum metal mirror, *c.* 1870s.

30-inch (76-cm) reflectors to promote extragalactic research, but the telescope had received little use; Woolley organized a programme of spectrographic research for it. He also took steps to refurbish the disastrous Great Melbourne Reflector as a 50-inch (127-cm) instrument, and he made plans to bring in a 74-inch (188-cm) reflector similar to that at the Radcliffe Observatory in South Africa. During World War II military needs took precedence; but, towards the end of 1945, the Radiophysics Laboratory demonstrated a portentous result: a correlation between radio noise at 200 MHz and the observatory's measurements of sunspot area. This led to an intense development of radio astronomy in Australia under the influence of such men as E.G. Bowen and J.L. Pawsey.

New Zealand

The fragmented nature of the history of astronomy in New Zealand results from the fact that the

European settlements came separately into half a dozen different areas cut off from mutual communication by severe geographical obstacles. Small observatories for navigational needs sprang up in these centres and have often grown into quite sophisticated amateur enterprises. The development of a limited degree of astronomical scientific research is a very recent phenomenon.

Time determination in the capital city of Wellington dates from 1862, being carried out from a variety of sites. Astronomy could not be a full-time occupation and was usually combined with seismology, land survey and museum curatorship, as in the case of J. Hector who directed the observatory at Kelburn as well as four other institutions. His successor, C.E. Adams, was also chief computer of the Lands and Surveys Department until 1914 when he became a full-time astronomer working on planetary and cometary ephemerides. In 1920 Adams became founding president of what

9.5. The old building of the Argentine National Observatory, Córdoba, 1872.

would become the Royal Astronomical Society of New Zealand. Soon after Adams's retirement in 1936 the observatory at Kelburn was officially assigned responsibility for general astronomical work and by reason of an 1896 bequest from C.R. Carter took Carter's name.

South America

Argentina

The Argentine National Observatory was founded through the vision and enthusiasm of two men: President Domingo Sarmiento, who was determined to raise his country from the depressed heritage of Spanish colonialism, and B.A. Gould (1824–96), who around 1865 had resolved to travel to the southern hemisphere to chart the southern stars with the detail achieved for the northern hemisphere. Gould had already selected Córdoba, in the Argentine hinterland 800 km by road northwest of Buenos Aires, as a suitable site relatively free from the coastal hurricanes of Argentina and the Andean earthquakes of Chile. Only later, when the Franco-Prussian war delayed the shipment of instruments from Europe and Gould turned his four North American assistants to a mapping of the sky by naked eye and with binoculars, did he appreciate the remarkable transparency of the atmosphere at Córdoba.

By the time of the observatory dedication in October 1871, Gould was able to say:

When, Gentlemen, after the Moon has set tonight, you raise your eyes to the starry sky, look carefully; and as

the smallest stars reveal themselves one after another, you will not find a single one whose place and magnitude has not already been recorded by some one or more of the astronomers of your observatory.

The printing of the atlas and text of the *Uranometria Argentina* was subjected to maddening delays, and not completed until 1879. This work clearly established 'Gould's belt' of bright stars, spread in a broad band inclined at some 20° to the galactic equator.

Although the municipal authorities at Córdoba spared no pains in helping the new observatory, the town was a rural outpost, and the astronomers always felt isolated. There were periodic political upheavals that cut off the observatory from the outside world, even from the printer in Buenos Aires. When deprived of government funds, as during four months in 1880, Gould paid the expenses out of his own pocket.

Gould returned to the USA in 1885, and was succeeded by the most productive of his original assistants, J.M. Thome (1843–1908). Thome had even greater monetary difficulties, particularly after the financial collapse of the Argentine government, and throughout the 1890s the observatory was supported by greatly depreciated non-convertible paper currency. The economic situation and constant threat of war with Chile prevented an increase in the paper currency allocation for several years. Despite this, Thome carried on with the observations for the two important catalogues, the great *Córdoba Durch-*

9.6. The southern station of Harvard College Observatory at Arequipa, Peru, 1892. El Misti rises in the background.

musterung, which is to the south what the *Bonner Durchmusterung* is to the northern hemisphere, and the Córdoba Zone *Astrographic Catalogues.*

Following Thome's death, the Lick Observatory astronomer C.D. Perrine (1868–1951) presided over a complete rehabilitation of the observatory. He brought to a close all the earlier programmes of positional astronomy, undertook pioneering southern galaxy studies, and laid the foundations for modern astrophysical research. Perrine's great ambition for the observatory was the figuring and mounting of a 60-inch (152-cm) mirror, but he failed to recognize the difficulties of undertaking such a project locally. Twenty years of fruitless effort on this telescope, combined with increasing hostility towards an essentially American outpost in the 'National' observatory of Argentina, finally forced Perrine to resign. Nevertheless, his efforts had paved the way for the eventual installation of a 60-inch reflector at Bosque Alegre in 1942.

Peru

The possibility of a southern-hemisphere station for the Harvard College Observatory was very much on E.C. Pickering's mind, when, in 1887, the bequest of U.A. Boyden became available to the observatory director. Boyden had left nearly $238000 in trust for the establishment of an observatory to be placed above the ill-effects of the earth's atmosphere. In 1889 Solon I. Bailey

(1854–1931) was sent to Peru in company with his photographer brother, Marshall Bailey, to set up a temporary high-altitude station. Initially based at Chosica, they later established the Boyden Station near Arequipa at 8055 feet. Living in tar-paper huts they produced in two years magnitudes of 8000 stars south of −30°, 2500 objective prism spectra to magnitude 8, and charts to magnitude 14.

In 1891 Pickering sent his brother W.H. Pickering on the second Peruvian expedition with instructions to lease land and to set up a permanent station. At Arequipa W.H. Pickering and his assistants installed a series of instruments including the 13-inch (33-cm) Boyden refractor and the 20-inch (51-cm) Common reflector. He purchased land and built permanent buildings and although these were to prove good investments, they almost bankrupted the Boyden fund, to the chagrin of the director at Harvard. Pickering was recalled, and Bailey again took charge.

Bailey initiated a period of great activity, restoring the dilapidated instruments, setting up a meteorological station on the nearby summit of El Misti, and beginning his celebrated studies of variable stars, especially those in globular clusters such as Omega Centauri and 47 Tucanae. In 1893 only 400 variable stars were known in the whole sky but this number rapidly increased as a result of Bailey's work. New discoveries included that of

9.7. The Large Magellanic Cloud. An exposure of almost 7 hours taken in November 1897 with the 8-inch Bache telescope at Harvard's southern station in Arequipa, Peru.

Nova Normae 1893 by Mrs Fleming at Harvard on an Arequipa plate. In 1894–95 a civil war broke out; though the observatory escaped damage, the optics were buried for a time as a precaution. In 1896 the Bruce 24-inch (61-cm) telescope arrived with its enormous 14×17-inch plates, each capable of photographing more than 25 square degrees of sky. By 1897, when Bailey returned to Harvard to produce his classification of variable star types, 293 variables had been discovered in the clusters. The division of work between Arequipa and Harvard was proving an effective arrangement; but among those at Arequipa some resentment arose from a belief that all the credit for discoveries was reserved for those at the parent observatory, as, for example, the discovery of Phoebe at Harvard by W.H. Pickering on an Arequipa plate of Saturn in 1899.

In 1904 the Bruce telescope began systematic observations of variables in the Magellanic Clouds, leading to the famous discovery of the period–luminosity relation (see Chapter 5). However, in the crucial Magellanic Cloud season, terrestrial clouds were prevalent at Arequipa and in 1908 Pickering sent Bailey to South Africa to prospect alternative sites. In 1909 Mrs Henry Draper, a financial pillar of Harvard astronomical activity, was compelled to curtail her support and Arequipa was the principal victim, the Superintendent, Leon Campbell, being left with only one helper. World War I forced closure in 1918 and when Paraskevopoulos reopened it after the war, the question of a new site was paramount, even though by 1924 the observatory had to its credit eight important papers on the Clouds and had even photographed northern sky sections accessible from Arequipa. After several South American sites had been investigated, Bloemfontein was preferred, and as has already been recounted, the Boyden Station transferred thither.

Further reading

David Gill, *A History and Description of the Royal Observatory, Cape of Good Hope* (London, 1913)

John E. Hodge, Charles Dillon Perrine and the transformation of the Argentine National Observatory, *Journal for the History of Astronomy*, vol. 8 (1977), 12–25

A.R. Hogg, Astronomical developments in Australia, *Journal of the Royal Astronomical Society of Canada*, vol. 47 (1953), 1–9

Bessie Zaban Jones and Lyle Gifford Boyd, *The Harvard College Observatory* (Cambridge, Mass., 1971)

Brian Warner, *Astronomers at the Royal Observatory, Cape of Good Hope* (Cape Town and Rotterdam, 1979)

Harley Wood, *Sydney Observatory, 1858 to 1958* (Sydney Observatory Paper no. 31, 1958)

10

Twentieth-century instrumentation

CHARLES FEHRENBACH

with a section on "Early rockets in astronomy" by Herbert Friedman

Introduction

The progress made in astronomical instrumentation in the first half of the twentieth century is
characterized by the extension of the spectral
domain and by the increase in sensitivity of
detectors of types already known or introduced
during this period. A second important development was the improvement of instruments and
their accessories, which led to far greater precision
in measurements of the positions, energy outputs
and spectra of stars.

The spectral range of the photographic plate was
widened so as to extend from the ultraviolet to the
near infrared, and the sensitivity of the emulsion
for a given grain size was markedly increased. The
vastly improved photographic plate remained an
essential tool because of its storage capacity, and it
also permitted utilization of wide fields (for example, with the Schmidt camera or with very high
dispersion spectra). Nevertheless, for precision
measurements and for the study of very faint
stars (except for astrometry), the photoelectric
cell and the ensuing electronic photography with
their associated techniques, superseded the photographic plate.

The progress made during this period is well
exemplified by the detailed photographic charting
of the heavens (leading to the National Geographic–
Palomar Observatory Sky Survey), by spectroscopic
studies at high dispersion and by the study of increasingly fainter objects. Measurements were also
made of the degree of polarization of light reflected
by the Moon and planets, and of the polarization of
the solar corona and of the stars. Meanwhile, for
most purposes visual observation was progressively
abandoned.

The new instruments included the Schmidt
camera, which allowed very large telescopes to
provide a wide field. The zenith tube and impersonal astrolabe made possible the measurement
of stellar positions with a precision never before
attained. Accessory devices were significantly improved in an important way: small prism spectrographs associated with refractors were replaced by
spectrographs with large gratings used with reflectors, yielding major gains in luminosity and resolution. The development of stellar and nebular
spectroscopy was certainly the most remarkable
aspect of astrophysics in the first half of the
twentieth century. In solar physics, the spectroheliograph, the underground Littrow grating
spectrograph, monochromatic filters and the coronagraph brought about considerable progress.

Photography

Photography, although widely used in astronomy
late in the last century, continued to grow in
importance in the first half of the twentieth
century. By 1950 the photographic plate had
become highly perfected, and, with its sensitivity
and wavelength response much increased, was
readily available from industrial laboratories.

Silver bromide emulsions in gelatine were sensitive only to radiation shorter than 5000 Å, but in
1877 the German photochemist H. W. Vogel, in the
course of depositing antihalation layers, noticed
that the dyes made his plates sensitive to red light.
At this time chemists were working intensively to
perfect new dyes, and in 1904 dicyanine made
possible panchromatic plates with a sensitivity to
7000 Å. Extension to the infrared was achieved
through other new dyes: in 1919 cryptocyanine
extended the sensitivity to 8200 Å, in 1925 neo-
cyanine extended it to 9100 Å and in 1934 a derivative of this dye extended the range to 13 500 Å.
(There is little hope of going further because of
fundamental physics linked to thermal noise in the

emulsion.)

Astronomers have actively collaborated with the photographic industry to improve the emulsions, but it is difficult to give precise dates for the use of the new materials in astronomy. Extensive research was undertaken by the industrial chemists such as C.E.K. Mees of Eastman Kodak to increase the sensitivity of the photographic plates and at the same time to increase their resolving power, but these two qualities cannot in general be achieved together. An important fault of plates is the so-called reciprocity failure, that is, the breakdown of the Bunsen–Roscoe law of reciprocity between the intensity of illumination and the exposure time (see Chapter 2). But I.S. Bowen's introduction of sensitization by heat under precise conditions did permit some improvement with respect to reciprocity failure.

The introduction of photography greatly influenced the design of optical instruments for astronomy. The older instruments, especially the refractors, were corrected for 5500 Å corresponding to the maximum sensitivity of the eye, and it was not possible to use them with unsensitized plates. Not until early in this century could the old refractors be used with yellow filters and panchromatic plates. By employing these refractors photographically, the determination of stellar positions, parallaxes, and proper motions became three or four times more precise than the visual micrometric measurements. This improvement was probably due to the fact that the plate recorded the mean image, thus minimizing the effect of turbulence.

For photography with plates that had not been colour sensitized, opticians constructed achromatic lenses corrected for the blue region of the spectrum. The older refractors had insufficiently wide fields, and to remedy this situation, increasingly complex lenses were calculated and constructed. Finally the Schmidt telescope provided the desiderata of a wide field up to 5° or even 50° together with negligible chromatic aberration and coma.

Photographs are taken at the focus of large parabolic reflectors, using a Ross corrector to compensate for coma. Numerous alternatives have been proposed to correct for coma and curvature of the field, and even put into effect in order to widen the usable field. Probably the best solution was that proposed by H. Chrétien and first carried out by G.W. Ritchey in the form of a 40-inch (102-cm) reflector built for the US Naval Observatory in the 1930s. However, the field of this coma-free combination of two mirrors is still more curved. In recent years special correcting lenses have been employed either at the prime focus or at the Ritchey–Chrétien focus to provide flat fields up to 1° and larger.

In spectrography, astronomers have used the photographic plate almost exclusively; not until after 1950 did the introduction of photoelectric receivers open a new avenue. Even so, the photographic plate has remained an indispensable detector for much astronomical research: in wide field photography for studies of galaxies and clusters, surveillance of the sky, astrometry and objective prism spectrography. It suffices to indicate that a plate obtained with the 48-inch (122-cm) Schmidt camera on Palomar has of the order of 4×10^8 image points. Not only are the magnificent photographs obtained using large telescopes (whether of Schmidt or classical design) remarkable sources of scientific information, but in the case of nebulae and galaxies their beauty makes them works of art appreciated by the public.

Electronography

The classical photoelectric tube allowed only the measurement of the flux of electrons emitted by the photocathode but failed to preserve the source image. Beginning in 1934 A. Lallemand (1904–78) developed at the Strasbourg Observatory an electronic camera that could refocus onto a photographic plate the electrons emitted by a photocathode, thereby obtaining an intensified image. The increase, originally of the order of ten, was extended to fifty for blue light and to much higher values in the red. This technique was difficult because it required the presence in the same high vacuum chamber of both a photocathode very sensitive to impurities and a photographic emulsion that *in vacuo* emits gases. Lallemand overcame these difficulties by cooling the plate; the photocathode was also cooled to minimize its thermal emission. Eventually Lallemand's instrument was developed into a number of different forms already foreshadowed in his original paper that appeared in vol. 203 of *Comptes Rendus de l'Académie des Sciences, Paris*:

10.1. The Lallemand electronic camera attached at the Newtonian focus of the Haute Provence Observatory 120-cm reflector.

The set-up that I assembled is as simple as possible, but it is quite likely that one could further increase the efficiency by receiving accelerated electrons not directly onto the photographic plate, but on a surface capable of multiplying the electrons; then one could again form the electronic image of this surface and utilize this image anew.

Electronic photometry

In the first half of this century, the earlier visual photometry and even photographic photometry were replaced by photoelectric techniques. Three main photoelectric effects had already been discovered in the last century: the photovoltaic effect (A.-C. Becquerel in 1839), photoconductivity (W. Smith in 1873) and the photoemissive effect (Hein-

rich Hertz in 1877). Nevertheless, their application to astronomy remained troublesome because of the weakness of the currents produced by the astronomical sources.

In 1907 Joel Stebbins (1878–1966) of the University of Illinois measured the flux of the Moon, and in 1910 he compared Algol to Alpha Persei, using a photoresistant selenium cell cooled by ice. The data were hard-earned because with a total deflection of 280 galvanometer divisions, Alpha Persei added only 7.25 divisions. A great advantage of the photoelectric measurements is their linearity, and from these data, Stebbins could readily deduce a magnitude difference of 0.18. These measurements, continued with great care, resulted in a beautiful light curve of Algol and the

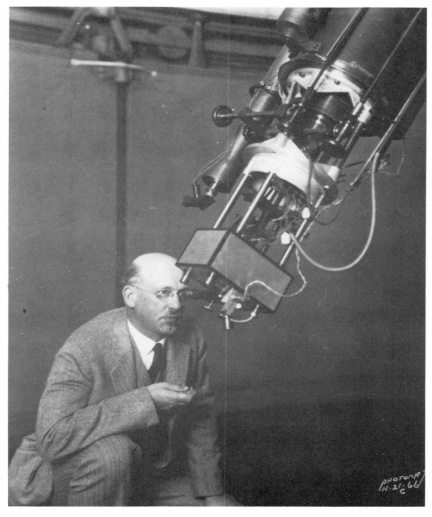

10.2. Joel Stebbins and his photoelectric photometer at the Washburn Observatory, *c.* 1929.

discovery of the secondary minimum, which had long been sought in vain with visual photometry.

In spite of this success, which was promptly repeated for Beta Lyrae, it soon became clear that photoconductivity failed to give precise measurements because of secondary effects, notably cell fatigue, and the same was true for devices depending on photovoltaic effects. After 1940 such cells were used in astronomy only for infrared applications. Thus, it was the photoelectric (photoemissive) cell, functioning in the range 3200–5000 Å and later extended to 12 000 Å, that found the principal use in astronomy, and astronomers adopted the new refinements successively introduced by the physicists.

Some alkali-metal surfaces emit electrons when struck by photons in a vacuum; a phenomenon that is at once simple and easily reproducible. But the electric currents produced, even by the best of present-day cells, are so small (75×10^{-6} amperes/lumen) that they are difficult to measure and often disturbed by parasitic causes. At first these currents were measured directly with a gold foil electrometer or with a string electrometer attached directly to the telescope. By 1912 P. Guthnick (1879–1947) had installed a photoelectric cell on the 30-cm refractor of the Berlin-Babelsberg Observatory; news of this development quickly reached Stebbins at Illinois, and the two groups published the first photoelectric results, on eclipsing binaries, almost simultaneously. But despite these early successes, in 1924 it was with difficulty that a 30-cm telescope could reach magnitude 7.5 for blue stars of spectral type A0.

Important progress was made along two different routes: the amplification of weak electronic currents and the extension of the spectral sensitivity. The amplification of currents was initially obtained with cells containing a residual atmosphere of rare gases. The gas molecules were ionized by primary photoelectrons, thereby considerably increasing the current. However, these effects were so difficult to control that precise measurement could not be undertaken, and thus the effect had technological applications only. It was amplification by the electronic tube (Lee de Forest in 1906) that marked the second stage in the development of photoelectric photometry. However, not until 1932 was this technique applied to the measurement of stars, by A.E. Whitford, who used a new type of vacuum tube developed especially for the amplification of small direct currents. At the same time, G. Rougier in Strasbourg applied it to measurements of the Moon.

This technique was rapidly replaced by the amplification of the photoelectrons themselves by means of secondary emission in a multiplier tube (V.K. Zworykin in 1936). With these photomultiplier tubes a multiplication of the order of 10^6 is regularly achieved. Among the first applications of photomultipliers in astronomy was the testing of an experimental Radio Corporation of America tube by Whitford in 1932 and G. Kron in 1937. After 1946 photomultipliers were used exclusively. For measuring the weakest astronomical sources, the counting of individual photons was also introduced from 1946 onwards, but digital techniques were not widely adopted for many years.

Improvement of the photoemissive surfaces took place rather slowly. From 1914 to 1940 a potassium surface sensitized by hydrogen was used. The cathode with multiple surfaces – caesium, caesium oxide – allowed a very large extension of the range of sensitivity, with a maximum in the near infrared. Complex surfaces of caesium and antimony, introduced in 1938 by P. Görlich, were three to four times more sensitive in the visual range than the caesium oxide surface. These improvements meant that the quantum efficiency of the photoelectric methods was perhaps twenty times higher than photographic emulsions.

By the 1950s the photoelectric photometer using a cooled photomultiplier tube had become a compact instrument. A well-defined colour filter facilitated the isolation of the spectral band that corresponded to the photometric system being used. The tube was often refrigerated, generally by dry ice, to reduce the 'dark noise'. Eyepieces allowed the observer to examine both the stellar field and the diaphragm, and the observer could move the diaphragm to measure the luminosity of the night sky. The time constant of the photometer was adjusted to the problem being studied: a small fraction of a second for rapid phenomena such as the occultation of stars by the Moon, or several seconds for the measurement of faint stars. The length of time for the recording also determined the time constant, and for the faintest objects it was sometimes necessary to record for several hours. In such a case a storage capacity of very high quality was required, but it was then possible to attain magnitude 21 or 22.

Thanks to photoelectricity, stellar photometry was enormously transformed in the course of the first half of the twentieth century. Visual and photographic magnitudes, poorly defined and imprecise, gave way to the UBV (ultraviolet-blue-visual) system of H. Johnson and W.W. Morgan (published in definitive form in 1953), in which the scale and zero point were accurately defined: a brightness ratio of 10 to 1 corresponded to a magnitude difference of +2.5, while six A0 Main Sequence stars with apparent magnitudes in the vicinity of 6 were specified as having the same apparent U, B and V magnitudes, and the zero point was fixed by a star assumed to be constant. The authors diligently measured several hundred stars, 10 of which served as primary standards and 280 as secondary standards.

The UBV system was actually in use a number of years before the definitive paper of 1953, so that by then Stebbins and Whitford had already extended this system up to 11 000 Å (six regions, U, B, V, G, R and I) for the brightest O and B stars. But the limited red sensitivity of their conventional caesium–caesium oxide cell restricted their measurements to the brightest stars. Extending the range to incorporate long wavelengths (up to 3 μm) became possible only around 1950 through the use of lead sulphide cells, so that the R and I magnitudes could be obtained for faint stars.

With the aid of photoelectricity, the atmospheric absorption could be determined with greater precision and the apparent magnitudes of stars, reduced to outside the Earth's atmosphere, became known to better than 0.01 magnitudes and perhaps nearly to 0.002 magnitudes. There is no doubt that the considerable improvement in photometry through photoelectricity was a revolutionary development for astronomy. Knowledge of the atmospheres of normal stars was greatly improved; interstellar absorption became measurable; and the photoelectric measurements of very faint stars led to a substantial revision of the magnitude scale for stars fainter than magnitude 19 and subsequently to a revision in the distance scale of the galaxies. Thus, by the early 1950s, photometry had at last become an exact science.

Electronic photometry of photographic plates
The first density measurements of stellar and spectrographic images on photographic plates were made with visual optical measuring machines by J. F. Hartmann at Potsdam in 1899. The human eye was soon replaced by photoelectric devices such as thermopiles, and later, photoelectric cells. As early as 1912 P. P. Koch designed a recording microphotometer so that the density of a spectrographic plate was immediately plotted against its displacement, thereby producing the first graphs of spectrograms. In the 1920s numerous such instruments were devised, one of the most widely adopted being the thermopile microphotometer invented by G. Moll at Utrecht. Other instruments were designed for estimating the density of a stellar image, and the method developed by J. Schilt in 1922 was used in many observatories. The introduction of the iris photometer in 1934 greatly increased the accuracy and reliability of magnitudes measured on photographic plates.

Bolometry

Attempts to measure the total energy flux received from a star as a whole were undertaken near the end of the nineteenth century. Unfortunately, this measurement turned out to be very difficult because of the insensitivity of those receptors that responded equally to all wavelengths.

The first experiments were made by E. F. Nichols

in 1898 using a radiometer modelled on a principle outlined by Sir William Crookes. The observed deflections were very small, and Nichols's results were never confirmed.

The first reproducible results were obtained by A. H. Pfund in 1913 and, especially, at the Bureau of Standards in Washington by W. W. Coblentz (1873–1962) in 1914. The instrument used in these two cases was an extremely tiny thermoelectric couple placed in a vacuum. In the case of Coblentz, the thermocouple was formed of bismuth and an alloy of bismuth with five per cent tin, and was 2 mm long and 67 μm wide. A small plate 335 μm in diameter was fastened to the couple. The sensitivity of the apparatus depended mainly upon the low heat capacity of the equipment and on its low thermal conductivity. Coblentz's instrument was 400 times more sensitive than those of his predecessors. With a very sensitive galvanometer, Arcturus gave a deflection of 85 mm and Vega 35 mm when the thermocouple was used at the focus of a 45-cm reflector. According to Coblentz, the difference in bolometric magnitude between Arcturus and Vega implied by these data is one magnitude. However, it became apparent that the measurements did not have a very precise physical significance on account of the very considerable infrared absorption of the Earth's atmosphere. Unlike the photometric measurements with relatively narrow, almost monochromatic bands, the bolometric magnitudes could not be corrected for the atmospheric absorption, so that the measurements of 110 stars published by Coblentz, while interesting to his contemporaries, did not stand the test of time. For this reason, attempts at the direct measurement of the flux energy of astronomical objects were not pursued with radiometers and bolometers after around 1920. Although the bolometer could be used for monochromatic measurements, its sensitivity was so weak that it was replaced by lead sulphide cells a thousand times more sensitive. And because of the very incomplete knowledge of atmospheric absorption, the flux emitted from very hot or very cool stars was known only roughly even in 1950.

Polarimetry

After the discovery of the polarization of light by reflection early in the nineteenth century, attempts were made to measure the polarization of the light from the Moon and the planets. But these observations, made visually with different polariscopes in France or in Britain, gave unreliable and even contradictory results.

In the 1920s B. Lyot (1897–1952) of Meudon Observatory finally succeeded in making excellent measurements of the polarization of the reflected light of the Moon and planets. At Meudon he built a visual polarimeter adapted from F. Savart's polariscope. This instrument consisted of two quartz plates cut at 45° to the optical axis. They were arranged one upon the other and followed by a Nicol prism or a tourmaline plate as analyser. Placed in front of the eyepiece of a telescope, the device produced fringes on the planet under observation.

In this decade Lyot's main progress stemmed from application of the principle whereby the light rays falling on the polariscope were gradually polarized alternately in a plane parallel to and in a plane normal to the principal section of the analyser. With this system fringes became visible, and by distinguishing two light contrasts in the fringes, one could detect the proportions of polarized light down to one per cent. Lyot then improved his instrument by adding a small known amount of polarization to the incoming beam. By slightly changing the angle of the added polarization he could, from the contrast of the fringes, detect an effect as small as 0.3%. These measurements, made between 1922 and 1925 and published in the *Annales* of the Meudon Observatory in 1929, remained the best published measurements of the polarization of the Moon and planets until 1950.

Lyot's visual polarimeter, mounted behind the prototype of his coronagraph, allowed him in 1930 to measure the polarization of the solar corona. In this way he was able in 1932 to show that the solar prominences were polarized between 0.3% and 0.12%. Meanwhile, in pioneering experiments beginning in 1923, Lyot tried to measure polarization photoelectrically. In his apparatus, the light being analysed was isolated by means of a small diaphragm placed in the focal plane of the telescope; it passed through a half-wave plate rotating rapidly in its own plane about its centre and then through a Nicol prism before being received by a photoelectric cell. The current of the cell possessed an alternating sinusoidal component whose amplitude was proportional to the quantity of light polarized, and this component was amplified, rectified and sent to a microammeter.

In 1924 Lyot replaced the Nicol prism by doubly refracting Iceland spar, the two beams being received into two cells. This doubled the available light, and the laboratory tests were satisfactory, but except for the Sun the results were not conclusive; the description contained in two sealed envelopes deposited with the Academy of Sciences in Paris in 1923 and 1924 were not published by Lyot until 1948, after he had had at his disposal photoelectric multipliers. Meanwhile, in 1942, Y. Öhman of the Stockholm Observatory had published the principle of an analogous polarimeter, but Öhman's instrument did not include a half-wave plate and it was the polarizer itself that turned. Because of this, the original instrument had a false zero, due to the polarization introduced by the cell, which was not the case in Lyot's instrument.

In 1946 H. W. Babcock at the Mt Wilson Observatory placed a differential analyser, consisting of a quarter-wave plate and a calcite crystal, in front of the slit of the coudé spectrograph of the 100-inch (254-cm) telescope and succeeded in measuring the polarization in the wings of some lines of 78 Virginis; this polarization, caused by the Zeeman effect, indicated the presence of strong magnetic fields. Within three years Babcock had measured variable magnetic fields of a few kilogauss in several peculiar A-type stars. The same general line of research led in the early 1950s at Mt Wilson to the solar magnetograph for mapping the magnetic fields of the Sun.

In 1947 the first stellar polarizations were measured at McDonald Observatory by W. A. Hiltner on the star RY Persei (for which S. Chandrasekhar had predicted a fairly high degree of polarization). This result was achieved photographically by passing the starlight through a Wollaston prism and a microscope objective, which formed two images of the telescope objective with orthogonal polarizations on a photographic plate. Hiltner found a degree of polarization scarcely larger than the error in his measurements.

Greater accuracy was later obtained by photoelectric techniques, using the 'flicker' method that was first published by Öhman and used by him in 1943 to measure the polarization of lunar surface features. In principle the polarized light from the

10.3. B. Lyot photographing the motions of the solar photosphere with a cinematographic camera attached to the 38-cm refractor at the Pic du Midi Observatory, May 1943.

source passes through a rapidly rotating analyser (such as a disc of Polaroid) and then falls upon a photomultiplier cell, the varying output of which is measured with an alternating current voltmeter. Rapid improvements in electronic methods enabled Hiltner and his colleagues to measure very small degrees of polarization from 1950 onwards.

The coronagraph

The solar corona, though easily visible during total eclipses of the Sun, is extremely weak. At 2′ from the edge of the Sun, its brightness is only a millionth that of the Sun's disc. This explains the failure from 1879 onwards of all the attempts to observe the corona outside of eclipse until success was finally attained in 1930 after the construction

of Lyot's coronagraph. His results showed that the previous efforts had revealed not the corona but only scattered light, a thousand times more intense than the corona.

The principle of the coronagraph, as conceived by Lyot, was identical to that of its predecessors: an objective projected the image of the Sun onto a disc of blackened bronze, situated in its focal plane and with a diameter some 10″ to 20″ larger than that of the Sun. After making a thorough study of the origin of the scattered light, Lyot understood how to overcome it. Part of it comes from the sunlight scattered by the Earth's atmosphere, minimized by observing at high altitude from a suitable site such as the Pic du Midi in France. Lyot showed that scattered light also originated from the objective:

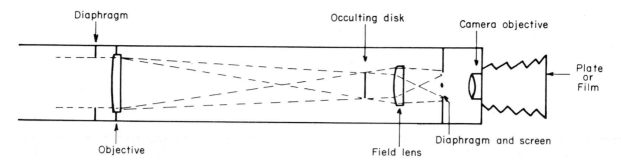

10.4. Schematic diagram of the Lyot coronagraph.

in general the glass contains bubbles and has irregularities of refractive index, and the lens surfaces are striated and contaminated by grains of dust. Finally, the edge of the objective diffracts light. For these optical investigations Lyot employed (and indeed invented) examination by the phase contrast.

To overcome these various nuisances, Lyot made the objective from a single piece of glass, carefully chosen for the absence of bubbles or irregularities and with its figure carefully ground. He specially cleaned the objective and took great care to exclude dust from the coronagraph. He eliminated the light diffracted by the edge of the objective by using a secondary objective to form an image of the principal objective on a diaphragm that has a circular opening a little smaller than this image (and a central reflection caused by the principal objective is also eliminated). The image of the corona was produced by a third objective. Of the objectives, only the principal one needed to be perfect.

In this way Lyot was able to trace the form of the solar corona by measuring the polarization of its light. In June 1938 Lyot also obtained excellent photographs of the corona out to several minutes of arc from the edge of the Sun and he was able to photograph the prominences.

It was possible to back up the coronagraph with numerous auxiliary instruments: spectroscopes to study coronal emission lines, a film camera to study the evolution of the prominences, a polarimeter to study the F corona. By 1950 coronagraphs were operating in many countries. They permitted a constant watch to be kept on the corona and the prominences, and on solar activity in general. But, of course, direct study during eclipses was as necessary as ever, especially for the observation of the outer parts of the corona.

Spectroheliographs and the Lyot filter

Spectroheliographs were developed in the 1890s by George Ellery Hale (1868–1938) in the United States and independently by H. A. Deslandres (1853–1948) in France. With this device they could observe the prominences both at the edge of the solar disc and in absorption on its surface, and they could examine the velocities of solar phenomena by setting the instrument's wavelength to the appropriate Doppler-shifted position. The instruments were solar spectrographs with mirrors and a grating or several prisms arranged to give a deviation of 180°. The image of the Sun was projected through the entrance slit and a monochromatic image selected by means of an adjacent exit slit. By moving the apparatus as a whole, the observer could slowly scan the disc of the Sun and obtain an image in monochromatic light on a photographic plate.

Numerous attempts were made to obtain successful monochromatic filters by other means, such as coloured glasses or gelatin filters, Christiansen filters and, more recently, interference filters. In 1927 Lyot published the principle of a polarizing filter and he put this into effect in 1933. In this instrument light passed through a series of polarizers whose planes of polarization were parallel. Between each polarizer was placed a crystalline strip made of quartz and cut parallel to the optical axis of the crystal. The faces of these strips were parallel to each other and at right angles to the light rays, and their optical axes were parallel and formed angles of 45° with the polarizer's planes of polarization.

Lyot demonstrated that the amplitude of the transmitted light is given by the classic formula for a diffraction grating with lines of $2''$; in the filter built by Lyot, $n = 6$. In general many transmission peaks are observed in the spectral domain of the receiver, but these can be separated by wide-band filters. The wavelengths of the peaks depend on the thickness and the temperature of the plates; it was easy to adjust them by controlling the temperature of the Lyot filter. Lyot actually constructed many versions of his monochromator, which differed according to the nature of the plates (quartz or Iceland spar) and the nature of the polarizers. In the best model the transparency achieved by Lyot was 40% in the visible and the equivalent width of the filter was 3 Å.

This filter, either in its original or derivative forms, is now universally used for the study of the solar chromosphere, and, in connection with a coronagraph, for the study of the corona. The coronagraph and the Lyot filter have given results of great importance for the study of the Sun and, because of their ability to monitor solar activity almost continuously, for solar–terrestrial relations.

The Schmidt telescope

The large refractors built at the beginning of the twentieth century gave good images only for small fields with diameters of a few minutes of arc. The majority of telescopes showed a residue of considerable chromatic aberration which made the images generally unsuitable for photography over an extended spectral range.

The invention of B. Schmidt (1879–1935) in 1931 of a camera combining reflecting and refracting elements giving a large achromatic and coma-free field represented a major step forward in optical astronomy.

Let us consider the principle of his invention. A spherical mirror forms at its focus an image blurred with an aberration that increases with the aperture of the mirror. This 'spherical aberration' is altogether eliminated if the mirror is made parabolic, but then the off-axis images are afflicted with strong coma. Schmidt corrected the spherical aberration of the spherical primary mirror by introducing an aspheric glass plate at least one surface of which was figured in just such a way as to compensate, zone by zone, for the spherical aberration of the primary mirror. The correction is

obtained by making the thickness of the corrector vary with the distance from the optical axis of the instrument. The corrector, which may be chosen from a sequence of possible forms, is placed at the centre of curvature for the mirror to provide for uniform correction over a large field. The thickness of the glass through which the light passes varies by $1/\cos \theta$, θ being the inclination of any given ray to the axis. Calculations show that Schmidt's solution leaves only a slight amount of aberration and that it is almost ideal. The correction for the spherical aberration is perfect only for a chosen wavelength, but the residual chromatic defects remain negligible over a large range of wavelengths around the mean.

Schmidt constructed the first astronomical telescope on this model in 1931 at the Bergedorf Observatory near Hamburg: its mirror had a diameter of 44 cm, its focal length was 62.5 cm and the corrector had a diameter of 36 cm. The telescope, which had a focal ratio of f/1.75, gave a field 15° in diameter (16 cm). The field was curved and the film had to be placed on a sphere of 62.5-cm radius. The corrector plate was difficult to figure, but Schmidt devised a solution to the problem. He placed the initially plane-parallel plate upon the end of a cylinder, which, when evacuated, deformed the plate. The upper surface of the plate, thus stretched and curved to the desired figure, was ground and polished to a flat surface. When the pressure was released the plate rebounded to the calculated aspheric figure. One can only admire the dexterity of this man, who had lost an arm in a boyhood accident.

This instrument was the first in a long series of Schmidt cameras that facilitated significant progress in astronomy. Most important was the Palomar Schmidt telescope with a focal ratio of f/2.44, a primary mirror of 72 inches (182 cm), a corrector nominally specified as 48 inches (actually 49.5 inches (126 cm)), a focal length of 120 inches (305 cm), and a plate area of 14×14 inches (36×36 cm) covering a field 6°.6 on each side. It first came into use in 1948. It followed an earlier Schmidt telescope on Palomar Mountain that has an 18-inch (46-cm) corrector and a focal ratio of f/2.

Figuring an optically precise aspheric surface was at first a difficulty. In 1935 Lyot and A. Arnulf showed that it was possible to replace Schmidt's

10.5. B. Schmidt in his optical shop at the Hamburg-Bergedorf Observatory, 1928.

corrector with a spherically surfaced doublet of null power but presenting exactly the correction necessary to compensate for the aberration of the spherical mirror. Independently, A. Bouwers (in the Netherlands) in 1942 and D.D. Maksutov (in the USSR) in 1944 suggested combinations in which Schmidt's corrector plate was replaced by a spherically surfaced meniscus lens.

In Bouwers's combination the three surfaces of the corrector and the mirror were concentric and this allowed a large field to be covered. Maksutov's system was more compact, but the field conditions were a little less satisfactory, although near achromatism with a simple element was achieved. Numerous other modifications of detail were introduced, notably in the construction of fully achromatic correctors.

Schmidt cameras are particularly well adapted to the photography of faint objects and transient phenomena such as meteors. To this end J.G. Baker (1914–   ) at Harvard in 1945 designed the 'Super Schmidt' camera. In this arrangement there are two concentric spherically surfaced meniscus elements or shells, one on either side of the centre of curvature of the spherical primary mirror. Within the shells there is also a 'hyperchromatic' doublet at the centre of curvature with at least one surface aspheric to a higher order of correction than for the classical Schmidt plate. Incident light passes through the first shell, through the doublet, and through the second shell; it reflects from the primary and then in converging passes back through the second shell a second time to come to an accessible focus just beyond. The focal ratio of the original instrument was f/0.66, but nevertheless a field of 52° diameter was successfully covered over an extended spectral range. The correcting system of the Super Schmidt camera was unusually complex and required a precision mounting. Although of only 12-inch (30-cm) aperture, the instrument weighed 2.5 tonnes. Several such instruments, mounted in pairs some kilometres apart in New Mexico, Hawaii and Canada, and photographing simultaneously, gathered considerable data on the trajectories of meteors.

Baker in 1940 also introduced several families of two-mirror and correcting-plate combinations with a flat focal plane accessible through a small central hole in the primary mirror, systems since known as 'Schmidt–Cassegrain'. These systems were further improved and analysed by E. H. Linfoot of Cambridge University. The first, and until 1950 the largest, such telescope was begun in 1948 and mounted at Bloemfontein, South Africa. This instrument has a corrector of 32-inch (81-cm) aperture and covers a circular plate of 8-inch (20-cm) diameter with sharply focused images.

Between 1935 and 1950 the literature contained a great many further suggestions for improvements to the Schmidt camera. In spectrography 'solid' Schmidt cameras introduced by D. Hendrix at Mt Wilson came into use. These specialized cameras are mentioned in the following section.

Spectrographs

Although the first photographic spectra were obtained by William Huggins and Henry Draper around 1876, little progress was made before the turn of the century. In the advance that then took place we can distinguish three phases: (1) the use of prism spectrographs on large refractors (1898–1910); (2) the construction of large reflectors especially adapted for spectrography, but always using prisms (1910–30); (3) the construction of large grating spectrographs with Schmidt cameras on large telescopes (since 1931).

Between 1898 and 1903 spectrographs came into service successively on the Lick (36-inch (91-cm)), Newall (25-inch (64-cm)), Potsdam (80-cm) and Yerkes (40-inch (102-cm)) refractors. Either three or four prisms giving a deviation of 180° were used. The prisms were small with heights of 3.2 to 5.2 cm, and the corresponding objectives were achromatic or anastigmatic. Various arrangements of the slit, the guiding eyepiece and the superposition of the standard spectrum were invented for these instruments and have become classic solutions. Numerous precautions were taken to eliminate flexure and to shield the instrument thermally.

However, there were several obstacles, including, at the beginning of the century, the difficulty of finding prisms of good optical quality. A fundamental problem was the chromatic aberration of large refractors made achromatic in the visible spectrum but retaining a large variation in focal length in the blue, which in turn led to numerous problems such as limited field, loss of light and errors in determining the positions of spectral lines. A small corrector lens for changing the achromatic correction from the visual to the photographic was used on the Lowell telescope, primarily intended for planetary spectroscopy; it obtained excellent results for the rotational velocity of Jupiter.

All these early spectrographs had very poor light transmission: when used in the blue there were large losses due to absorption in the telescope lens and the spectrograph optics, and above all within the dispersive glasses needed for the prisms. In addition, the losses due to reflection by the surfaces of the prisms were considerable. Furthermore, these early spectrographs employed rather slow camera objectives that in turn required a very narrow entrance slit for satisfactory resolution. Indeed, the width of the entrance slit was comparable to the element of resolution of the photographic plate (of the order of 0.03 mm). The angular width of such a slit projected onto the sky was generally much less than 1 arc second (and in fact, around 0.25 arc second for most of these spectrographs). Only a small fraction of the starlight then entered the slit. The optical transmission, found by taking into account the losses in the optics, the narrow slit-width and the lack of full achromatism, was around 1% according to several authors. It is scarcely surprising that the Yerkes spectrograph needed a two-hour exposure on a 5.5 magnitude star with 0.5 arc second image and a dispersion (or, more strictly, an inverse dispersion) of 13.5 Å mm^{-1} at 4500 Å. The Newall spectrograph at Cambridge, England, reached fourth magnitude in 80 minutes with a dispersion of 25 Å mm^{-1}. These spectrographs achieved the first reliable determinations of radial velocity; the Lick results obtained by W. W. Campbell and J. H. Moore were considered excellent and were long used to provide standards for calibrating other instruments, whereas the Potsdam ones were less successful.

The faint diffuse images of nebulae posed a particular problem and required fast, short-focus camera lenses in the spectrograph; with such a combination V. M. Slipher of the Lowell Observatory had determined the radial velocities of 41

10.6. W. W. Campbell standing alongside the original spectroscope on the 36-inch Lick refractor. This spectroscope, constructed by J. A. Brashear, was designed and used by J. E. Keeler. The picture was taken in the summer of 1890 when Campbell was a volunteer assistant to Keeler at Lick.

galaxies by 1925. The discovery made by E.P. Hubble in 1928, of a velocity–distance relation for the extragalatic nebulae, and the urgent need to extend the studies to still fainter and more distant objects, required a further development. A number of high-speed objectives of large aperture were designed. They were later replaced by small, fast Schmidt cameras made effectively of a single block of glass with only one air–glass interface at the entrance aperture, the 'solid Schmidt'. It is impossible to list the many other spectrographs, but we must mention the one on the 36-inch (91-cm) Crossley reflector, which had two quartz prisms 5 cm high and two quartz lenses in an arrangement proposed by J.F. Hartmann in 1900. When used in a slitless configuration proposed by F.L. Wadsworth, there was almost no spherical or chromatic aberration.

At the other extreme, the first high-dispersion spectrograph was that constructed for the coudé focus of the Mt Wilson 60-inch (152-cm) under the direction of W.S. Adams. This instrument contained a 63° flint prism backed by a plane mirror that caused the beam to return through the prism, thus doubling the dispersion. The focal length of the autocollimating lens was 5.5 m and it gave a dispersion of 1.4 Å mm^{-1} at 4300 Å and 6.2 Å at mm^{-1} 6500 Å. Regrettably the prism, which had a face only 12.7 cm across, was too small and more than half the light was lost by vignetting. Only spectra of the brightest stars were secured and measured: Sirius, Procyon, Arcturus, Betelgeuse, Rigel and Antares. These spectra were, however, of high quality and they allowed Adams to study variations in radial velocities as a function of the ionization state in a stellar atmosphere.

As early as 1912 Hale at Mt Wilson had recognized that it was imperative to provide a new source of large, high-quality gratings for the large solar and stellar spectrographs that were being constructed. He established a laboratory for this purpose and brought to Pasadena J.A. Anderson, an experimental physicist who had rebuilt H.A. Rowland's ruling engine at Johns Hopkins University. The ruling engine built by Anderson at the Mt Wilson Observatory produced half a dozen gratings. In 1929 the design of an improved ruling engine, smaller and much more rigid, was begun by H.D. Babcock. Around 1931 John Strong, working in the Mt Wilson Observatory shops,

introduced the vacuum-evaporation process for aluminizing telescope mirrors; with his collaboration, Babcock found that thick (1 μm) aluminium films deposited on glass were vastly superior to speculum metal for gratings. Almost overnight it became possible to rule blazed gratings with grooves of a sloping profile, concentrating 70–80% of the light into a single chosen order. This innovation was also adopted by R.W. Wood at Johns Hopkins University, where he was using the rebuilt Rowland engine.

The spectacular evolution of the stellar spectrograph in the late 1940s was triggered by the increasing availability of blazed gratings ruled with greatly improved precision, and by the adaptation of the Schmidt camera to the spectrograph, already initiated by Adams and T. Dunham at Mt Wilson in the mid-1930s. With three to four times the dispersion of prisms, far better resolving power, and with a reflectivity markedly superior to the transmissivity of prisms over a wide spectral range, blazed gratings could be readily combined with camera objectives three to four times faster than before, all for the same final dispersion. This combination resulted in the angular width of the slit on the sky becoming correspondingly larger. These newer spectrographs now admitted ten to twenty times as much light as the old prism spectrographs, which now are of use only for low-dispersion ultraviolet spectroscopy. Because of this, practically all spectrographs made since 1948 employ diffraction gratings.

Spectrographs at the primary or Cassegrain focus generally use gratings with a ruled height from 5 to 10 cm. It can be shown that the light grasp is proportional to the height, which explains why ever larger gratings are in demand, up to 30 cm for a composite quadruple grating for the Hale telescope. Naturally, these are bulky and heavy spectrographs that can be placed only at the coudé focus. At this fixed focus they benefit substantially from the mechanical and thermal stability of the enclosed coudé room. Results obtained with such instruments have led to immense progress in astrophysics, more especially as the spectral range covered is from the ultraviolet (3200 Å) to the near infrared (12000 Å).

Objective prisms

Slit spectrographs are excellent instruments for the

study of isolated stars, but their transmission (of the order of 1% at the beginning of this century) was poor and it was possible to obtain the spectrum of only one star at a time. It therefore occurred to astronomers as early as 1890 to place a prism of relatively small angle (of the order of 10°) in front of an astronomical objective to replace the image of each star in the field by its spectrum.

The most systematic work on this was carried out at Harvard College Observatory by a team who used two objective prisms, one of which was placed on the 8-inch (20-cm) Draper telescope and the other on the Bache telescope at Arequipa, Peru. Originally a pairing of prisms of 13° and 5° respectively was envisaged for the two instruments, whose relative rotation would allow the dispersion to be varied. The length of the spectra between Hβ and Hε was from 5.6 to 1.6 mm at Harvard and from 5.8 to 2.2 mm at Arequipa. The corresponding inverse dispersions were of the order of 150 and 520 Å mm^{-1} at Hγ. In fact, the first prisms were not very satisfactory and new ones were constructed, which gave the same dispersion as the original prisms. Higher dispersions were adopted for brighter stars. These instruments made possible the compilation of the monumental *Henry Draper Catalogue*, which gave the spectra of more than 225 000 stars, while later additions brought the total to nearly 300 000.

In the first thirty years of this century numerous other objective prisms were used in various observatories. The Lippert astrograph at the Hamburg Observatory had an objective 34 cm in diameter and a focal length of 340 cm. It was fitted with a dense flint-glass prism of 8.5° with a clear aperture of 30 cm (in 1912), and also with a 9.5° crown-glass prism (in 1923). The corresponding dispersions were 100 Å mm^{-1} at Hγ (faintest magnitude 9). The same prisms used with an astrographic triplet yielded a dispersion of 230 Å mm^{-1} towards Hγ (faintest magnitude 10). It was in this way that the spectral classification of the stars of the 115 Kapteyn fields was obtained.

There was, however, one major disadvantage common to such instruments: the variation of the focal length with the wavelength limited the spectral range, and in addition, the angular field of the objectives was always small. It was therefore a significant advance when, in 1939, objective prisms were first coupled with Schmidt telescopes. The 24-inch (61-cm) Burrell telescope at the Warner and Swasey Observatory had a 4° prism of the same aperture. Its dispersion was 283 Å mm^{-1} at Hγ and it could reach magnitude 12 in a 30-minute exposure with a 5° field. An identical instrument was installed around 1949 at the University of Michigan. A Schmidt telescope at Tonantzintla in Mexico was equipped in 1939 with a 4° prism and a clear diameter of 26 inches (66 cm), giving a dispersion of 250 Å mm^{-1} at Hγ and reaching magnitude 13.3 in a 90-minute exposure. These Schmidt telescopes in combination with carefully figured prisms gave spectra far superior to those obtained with ordinary refractors.

These objective prisms nevertheless provided plates suitable only for spectral classification and analysis. A zero-deviation objective prism with normal field intended for the measurement of radial velocities *en masse* was first devised and constructed by C. Fehrenbach (1914–   ) of the Haute Provence Observatory. The prototype comprised a barium crown-glass prism to which were coupled on each side flint-glass prisms whose angles were half as wide as the angle of the barium crown prism so as to constitute a unit with parallel faces. The glasses were carefully chosen to have the same index of refraction at a mean wavelength near 4220 Å, but the crown and flint dispersions were widely different. Because the instrument was a direct vision device over the entire field at the mean wavelength, it had the particular advantage that one could employ directly the method proposed by K. Schwarzschild with a 180° rotation between exposures, but which had, up till then, proved impractical to use because of the differential corrections over the field. The first version of this objective prism, 15 cm in diameter, came into use in 1946. It reached magnitude 9.5 with a dispersion of 80 Å mm^{-1} between Hγ and Hδ. Subsequent models, 40 to 60 cm in diameter were made of only two prisms with the same angle, one of flint and the other of crown barium but connected together to form a plate with parallel faces. The precision of the individual measurements of radial velocities was not comparable to that achieved with slit spectrographs, but it nevertheless permitted excellent statistical studies of faint stars with an accuracy of the order of ± 5 km/s.

10.7. J.J. Nassau attaching the objective prism to the Burrell Schmidt at the Warner and Swasey Observatory, Cleveland; in the background is A.P. Leary.

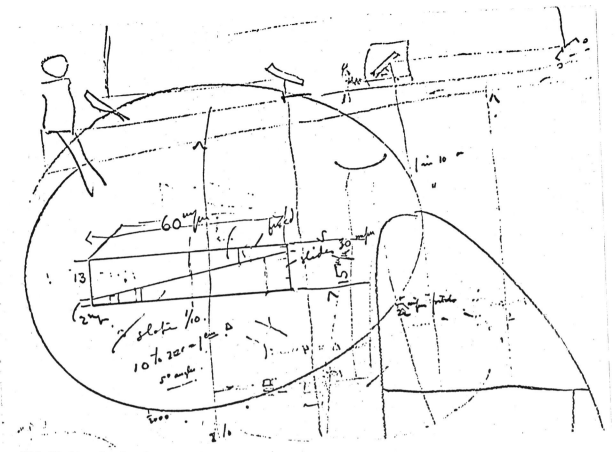

10.8. Working sketches from a notebook of F. G. Pease made in July 1920 as the Michelson interferometer was being prepared in the Mt Wilson shops. Before the mirrors were controllable from the Cassegrain focus of the 100-inch, an assistant had to sit on top of the beam, as illustrated, to make fine adjustments. Note also the design of the optical wedge needed to equalize optical path lengths.

Michelson's interferometer for measuring stellar diameters

When two circular apertures of diameter d, whose centres lie at a distance D from each other, are placed in front of a telescope mirror, a diffraction disc surrounded by rings is observed for a stellar point source. The radius of the first dark ring is $\rho = 1.22\lambda/d$, where λ is the wavelength of the light. The whole of the diffraction pattern is crossed by dark and light fringes. The distance between the fringes is given by $\delta = \lambda/D$. If the star is double with a separation of $\delta/2$, the bright fringes of the one star cover the dark fringes of the other, which then disappear. This principle was pointed out by A.-H.-L. Fizeau in 1868 and applied by E. Stephan at the 80-cm telescope of the Marseilles Observatory in 1873–74; his observations indicated that the bright stars should have angular diameters much smaller than 0″.158.

It would seem, *a priori*, that atmospheric turbulence, which acts differently on the two beams, should obliterate the fringes. However, A. A. Michelson (1852–1931) showed that this was not at all the case: the fringes move but do not disappear, and, as long as the period is not too short and the amplitude not too large, their movement can be seen and followed. In 1891 he used this method with the Lick 12-inch (30-cm) refractor to measure the diameters of the Galilean satellites of Jupiter.

Using a calculation that takes into account the limb-darkening of the star, one can deduce the

10.9. The Michelson interferometer on the 100-inch telescope, 10 August 1920.

diameter of the star from the distance D between the branches of the interferometer for which the fringes disappear. But Michelson did not at first follow up on the interferometric procedure, because the distance D appeared to be several tens of metres, impractically large. Not until Ejnar Hertzsprung and Henry Norris Russell developed the concept of giant stars did the work resume. Then it was realized that for the largest stars, D, could be as small as several metres.

In 1920 Michelson designed a large interferometer and it was mounted on the 100-inch (254-cm) telescope at Mt Wilson by F. G. Pease and Anderson. This instrument had a 20-ft (6-m) steel beam, carefully designed to reduce the flexure to a minimum. On this beam were mounted two movable mirrors whose spacing could vary between 2 and 5.5 m. If there were no flexure, one would only need to change the distance D between these mirrors to be able to see the fringes appear and disappear and hence to deduce the diameter of the star (the limb-darkening of the star being known). Michelson finally succeeded in contriving very efficient optical devices to compensate for the mechanical flexures.

When the telescope was pointed at Betelgeuse on 13 December 1920, the interference fringes disappeared on the principal image for $D = 252$ cm. When pointed at Algol and Gamma Orionis, the fringes were visible on both images, which proved that the adjustment was still accurate and the disappearance not due to any defect. The diameter of Betelgeuse was calculated from this experiment as $0''.047$ for a uniformly lit disc, or $0''.055$ if a certain amount of limb-darkening was allowed.

The first success was quickly followed by measures of other red giants. At first Michelson maintained an active interest in further refinements, but eventually ceased participating. Pease persevered well into the next decade, designing and using an independently mounted 50-ft (15-m) interferometer, but this device never fulfilled his expectations, and he finally gave up in disgust.

The zenith telescope and the prism-astrolabe

The meridian transit telescope, which was introduced in the eighteenth century, has been used over the years to provide very precise coordinates for numerous stars and so establish a positional reference frame over the sky. Such telescopes also provided the precise measurement of time. They also supplied evidence for irregularities in the Earth's rotation (see Chapter 13), and from early in this century it was realized that the study of the details of the Earth's rotation and the variations in the position of the celestial pole was a subject of intrinsic interest, quite apart from the question of improving the measured stellar declinations.

It is atmospheric refraction that limits the precision of declination measurements. Two instruments have been devised to minimize this effect: the zenith telescope and the impersonal astrolabe. Experience has shown that both these relatively simple and stable instruments can also be used to find the right ascensions of certain stars and to determine time with precision.

The zenith telescope

As early as 1851, G. B. Airy installed at Greenwich a zenith instrument, which was used visually until 1910. However, only at the beginning of the present century was a high precision instrument of this type developed, by F. E. Ross (1874–1960). Its success was then due to an ingenious placement of components and to the use of photography, from which the name PZT (photographic zenith tube) derives.

The essential part of Ross's instrument was a special astrographic objective, about 20 cm in diameter, with a focal length just over 5 m. The axis was directed to the zenith. A mercury bath was placed about half a focal length beneath to define the vertical. With this arrangement a star visible at the zenith gave an image situated just under the objective, where it could be recorded on a photographic plate.

Ross chose to use a moving plate in order to record fainter stars. He installed his instrument at Gaithersburg, Maryland, and between June 1911

and October 1914 he photographed 6944 stars on 450 nights. The instrument was subsequently acquired by the US Naval Observatory, where it remained in continuous service as 'PZT no. 1' from 1915 to 1955. During this time further improvements were effected: for example, a timing contact was added so that from 1934 it could be used for the determination of time.

The photographic zenith tube has only a small field, of the order of 30', but even so a significant number of stars brighter than magnitude 8.5 could be photographed; of course only stars passing close to the zenith of the observatory, in a narrow band some 30' wide, could be observed. A certain number of star groups, say eight, were observed during the course of the year. By comparing these groups one could considerably improve the coordinates of the stars and study the variations in the position of the pole or in the latitude of the observatory. Additional instruments of this type have been set up, for example, at Richmond (Florida), Greenwich and Tokyo.

The Danjon astrolabe

The prism-astrolabe of F. A. Claude and L. Driencourt (1905) had a 60° prism whose apical edges were aligned horizontally: situated behind this was a telescope whose optical axis was horizontal and perpendicular to one face of the prism. If a mercury pool was correctly placed, it was possible to observe two images of a star that was situated at 30° from the apparent zenith (see Figure 10.11). These two images travelled in opposite directions and coincided at the moment when their zenith distance was exactly equal to 30° (that is, for a prism whose angle was exactly 60°). This device, very simple in principle, was intended as a field instrument for determining geographic latitudes. But it was transformed by André Danjon (1890–1967) into a high precision astronomical instrument.

The chief difficulty of the initial instrument lay in recombining the beams, because any focusing error displaced the relative position of the two stars, thus making it impossible to obtain precision measurements. Danjon placed a Wollaston double prism in the beam of the convergent rays to give four beams of polarized light, two for each of the two incoming beams. The two central beams, each emerging from one of the pairs, were parallel and

10.10. The Ross photographic zenith tube of the US Naval Observatory with observer S. Edelson, *c*. 1950.

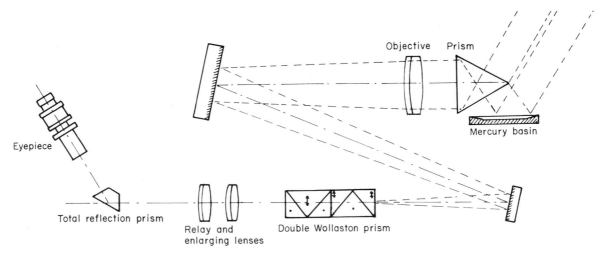

Objective   Prism

Mercury basin

Eyepiece

Total reflection prism   Relay and
enlarging lenses   Double Wollaston prism

10.11. Schematic diagram of the Danjon prism-astrolabe.

the deleterious effects of focusing thus eliminated. The instrument was then made impersonal: all that was needed was to set the prism with a theoretical displacement (depending on the latitude of the observatory and the azimuth of the star) to maintain the coincidence of the two stars. In fact, the stars are placed side by side and the observer keeps them exactly between two parallel wires. It is the correcting movement which is recorded and which constitutes the measurement.

The great advantages of the prism-astrolabe were that it observes a far larger zone of the sky than the zenith tube and that the angle of the prism serves as a fixed parameter that does not vary from summer to winter (in comparison to the meridian telescope, which depends upon the seasonal stability of the pivots and the collimators).

The zenith tube and, more particularly, the Danjon astrolabe, have led to small but necessary changes in the fundamental catalogues compiled with meridian instruments. It has been established in this way that the differences of right ascension (Astrolabe − FK3) are positive for stars situated around 60°N and very slightly positive for stars situated around 30°. In right ascension the correction can reach +0″.10, owing to systematic errors. These instruments have made possible a basic improvement in the fundamental constants of astronomy.

Early rockets in astronomy

Three men, K.E. Tsiolkovsky (1857–1935), R.H. Goddard (1882–1945) and H. Oberth (1894–   ), developed the basis of modern rocketry early in this century. Tsiolkovsky, a Russian high-school teacher in the town of Kaluga, derived mathematically the basic relationship between the terminal velocity of a rocket, its exhaust velocity and the ratio of mass of rocket to mass of fuel. Goddard not only theorized about rocket flight, he designed and tested his concepts over a period of thirty years. He launched the first liquid fuel rocket in Auburn, Massachusetts in 1926, and over the following two decades Goddard engineered many of the components of modern rocketry and conducted the first scientific measurements at high altitudes. Oberth covered much of the same theoretical ground as Tsiolkovsky and organized a 'Society for Space Flight' whose membership included the youthful Wernher von Braun. Years later, von Braun led the development of the V-2 rocket for military purposes; thousands of V-2's were aimed at London and Antwerp in 1944 and 1945.

With the end of World War II, US Army forces captured a stockpile of V-2 rockets in Germany. In January 1946 plans were formulated to use some twenty-five V-2s for scientific experiments. Following initial successes, the programme was expanded

10.12. The V-2 rocket in the gantry at the White Sands Missile Range in New Mexico, 1948. The V-2 was 14 m tall and designed to carry a 900 kg payload to as high as 150 km.

until more than seventy-five rockets were available in the years to the end of 1950. The warhead, designed for 900 kg of explosive, was instead filled with scientific instruments. In contrast to Goddard's last rocket, the V-2 was enormous. It measured 14 m long, 1.7 m in diameter, and weighed 14 tonnes when loaded. A typical successful flight from the White Sands Missile Range in New Mexico would last about 6 minutes and attain an altitude of 120 km. The rocket was spin-stabilized, but received some control during powered flight from graphite steering vanes in the jet.

During free fall, the V-2 reached speeds of about 1 km per second and impacted with such force that it almost totally disintegrated. Techniques were quickly developed to break up streamlined flight, such as separation of the warhead with explosive charges at 50 km on descent. The rocket parts then fluttered down at speeds less then 0.1 km per second. With such precautions to soften the impact, recovery of scientific instruments became possible. At the same time, radio telemetry techniques were developed to transmit scientific data throughout flight. Towards the end of the 1950s, parachute techniques also came into use for recovery of scientific payloads.

The early scientific programme was a mixture of upper atmosphere research, ionospheric probing, cosmic ray measurements, and solar ultraviolet and X-ray spectroscopy. Extension of the ultraviolet spectrum to wavelengths shorter than the ozone cut-off at 3000 Å was undertaken by research teams at the Naval Research Laboratory (NRL) led by E. Durand and R. Tousey and at the Johns Hopkins Applied Physics Laboratory by J. J. Hopfield, H. E. Clearman and J. A. Van Allen. The NRL group achieved the first success on 10 October 1946. Their spectrograph was fixed to the rocket without provision for any controlled pointing toward the Sun. In order to maximize the exposure, conventional entrance slits were replaced with 2-mm spheres of lithium fluoride. Tiny images of the Sun formed behind the beads and permitted useful spectroscopic exposure for all solar angles up to 70°. Wavelengths down to 2100 Å were registered, although spectral lines were badly smeared.

It was clear that pointing controls were essential. Single axis systems were quickly adopted by the Applied Physics Laboratory and the NRL. The Applied Physics Laboratory group had success soon afterwards and the NRL effort, under Tousey's long-term guidance, steadily improved the quality of the spectra and extended the range deeper into the ultraviolet. The change to two-axis stabilization took almost 6 years and was accomplished by the University of Colorado under contract from the US Air Force. Once available for general use, this stabilizer permitted extension of the solar spectrum all the way to its X-ray limit.

Broad-band photometric measurements were made by H. Friedman of the NRL in 1949. The wavelength intervals covered were 1–8 Å, 1100–1350 Å and 1450–1750 Å and all data were telemetered continuously. The Lyman-α line of hydrogen was detected at a height of 75 km, followed by soft X-rays above 85 km. The Schumann continuum of the ultraviolet increased steadily from 90 km. In this first successful broad look at much of the short-wavelength solar spectrum, H-Lyman-α was identified as the source of the ionospheric D-region and soft X-rays as the source of E-region. The absorption of Schumann ultraviolet revealed the extension of molecular oxygen to altitudes far above the level predicted by photochemical equilibrium theory.

These observations set the stage for rocket astronomy up to the Sputnik era. Within those few years from 1949 to 1957, high-resolution solar spectra were obtained at all wavelengths and the high variability of the extreme ultraviolet and X-ray emissions was revealed. Most spectacular of all was the X-ray flash of a solar flare, first observed by Friedman and his colleagues in 1956. Galactic astronomy was also initiated in 1956 by the NRL group with observations of early type stars in the far ultraviolet and the first hint of detection of X-rays from beyond the solar system.

Conclusion

While the first half of the twentieth century saw the building of ever larger reflectors, the gains made by the advances in auxiliary instrumentation and detectors (including photography) more than equalled the increased light-gathering power of the greater apertures. In addition, a wider spectral range permitted exploration in the ultraviolet and near infrared, and, as the next chapter indicates, in the radio region as well. In 1950 astronomy stood on the threshold of an even wider exploration of the previously inaccessible spectral domains. The host of new instrumental devices and carriers – a veritable electronic revolution – will be mentioned in the Retrospective Essay.

Further reading

André Danjon and André Couder, *Lunettes et Télescopes* (rev. edn, Paris, 1979)

Zdeněk Kopal (ed.), *Astronomical Optics and Related Subjects* (Amsterdam, 1956)

Gerard P. Kuiper and Barbara M. Middlehurst (eds), *Stars and Stellar Systems*, vol. 1: *Telescopes* (ed. by Kuiper and Middlehurst, Chicago, 1960), and vol. 2: *Astronomical Techniques* (ed. by W. A. Hiltner, Chicago, 1962)

C.E. Kenneth Mees, *The Theory of the Photographic Process* (New York, 1946)

G.R. Miczaika and William M. Sinton, *Tools of the Astronomer* (Cambridge, Mass., 1961)

V.K. Zworykin and E.G. Ramberg, *Photoelectricity and its Applications* (New York, 1949)

11

Early radio astronomy

WOODRUFF T. SULLIVAN, III

When the history of astronomy in the second half of the twentieth century comes to be written, a place of outstanding prominence will be accorded to radio astronomy. The achievements of the radio astronomers in the first half of the century are therefore of unusual historical significance, transcending their impact within that period. Since radio astronomy grew from a symbiosis of radiophysics and electrical engineering with little help from astronomers, we shall place strong emphasis on the instrumentation and techniques; for in fact it was not until the late 1950s that integration into astronomy as a whole was well under way. Space limitations have not allowed any discussion of early *radar* astronomy, concerned with the study of reflections from the Moon and from ionized meteor trails.

Galactic background radiation

Despite the enormous growth of radio technology in the first few decades of this century, there were no purposeful attempts to detect emission at radio wavelengths from the Sun or other extraterrestrial objects between 1902 (of which more shortly) and 1932. The first discovery, in 1932, in fact was accidental; it was made by an American radio engineer, Karl G. Jansky (1905–50), who was studying the nature of interference occurring on the recently inaugurated trans-Atlantic radio telephone service. Despite his strong interest in the extraterrestrial 'star static' and attempts to interest radio engineers, astronomers and physicists, Jansky's work was little appreciated. The radio engineers and astronomers did not know enough of each other's discipline to recognize the potential of this discovery; the world of decibels and superheterodyne receivers was simply too far removed from that of spectroscopic binaries and proper motions.

There were four keys to Jansky's discovery: (1) The great sensitivity and stability of his receiver in the newly-exploited shortwave range (20 MHz frequency or 15 m wavelength) where ionospheric effects were less dominant, (2) the relatively large directivity ($24° \times 36°$) and steerability of his 30-m long antenna, (3) the circumstance that 1932 was a year of solar minimum, allowing far easier recognition of the extraterrestrial origin of the signals since the ionosphere was less active and less absorptive, and (4) Jansky's curiosity and drive to understand the origin of the 'steady, weak, hiss-type static', despite the fact that its importance to communications was minimal at that time. From one year's worth of data of the sort shown in Figure 11.1, he established that the radio waves were extraterrestrial and associated with the band of the Milky Way, the strongest signals coming from the direction of the galactic centre in Sagittarius. He could only speculate on their source, noting that an interstellar origin seemed more likely than stellar, especially as he had not detected any radio emission from the Sun. Jansky published his findings in the *Proceedings of the Institute of Radio Engineers* between 1932 and 1935, and soon moved on to other aspects of radio noise research at Bell Telephone Laboratories.

In the ensuing decade, only one man, Grote Reber (1911–), recognizing the importance of this extraterrestrial radiation, carried out extensive investigations as best he could with limited resources. During the day, Reber was a radio engineer in Chicago; at night he was a radio astronomer in his backyard in the suburb of Wheaton, Illinois. In 1937 he designed and built a 9-m paraboloidal reflector illuminated by a novel feed system. The magnitude of his engineering skill and ingenuity can be appreciated only by realizing that this part-time, backyard project produced

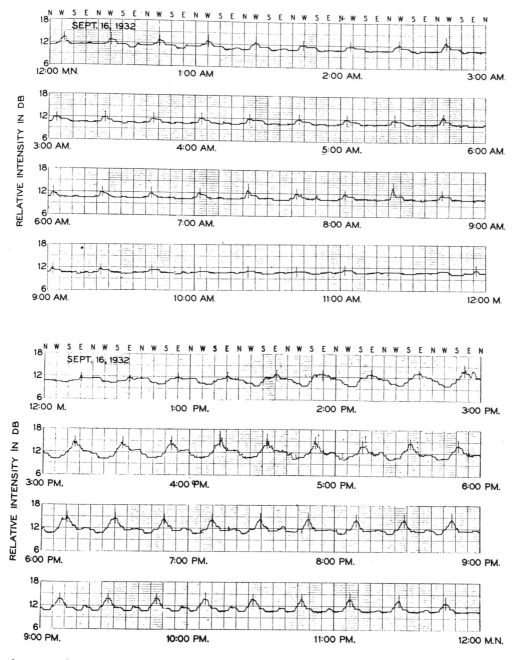

11.1. The discovery of extraterrestrial radio radiation. This strip chart recording of 16 September 1932 is all that remains of Jansky's original year's worth of 20.5-MHz data taken at Holmdel, New Jersey. Depending on the orientation of the Milky Way with respect to the local horizon, each 20-minute rotation of the antenna yielded either one or two prominent humps.

what was undoubtedly the largest antenna of its kind in the world. His 'dish' not only provided angular resolution much improved over Jansky's antenna, but allowed operation at considerably higher frequencies where, Reber reasoned, the radiation should be more intense if it followed a blackbody law. Despite unsuccessful attempts at 3300 and 910 MHz in 1938, Reber persisted, and in 1939, at 160 MHz, he was at last able to confirm Jansky's discovery.

Reber's four primary papers, published between 1940 and 1948 and all entitled "Cosmic static", describe six years of mapping the galactic background radiation at 160 and 480 MHz, always as a part-time activity. Yet his contour maps were not improved upon for almost a decade, and they clearly delineated the Milky Way as well as prominent regions of higher intensity that were only later recognized as discrete sources (Cas A, Cyg A, Sgr A, Vela X, etc.). The only other complete survey at this time was that in England by J.S. Hey (1909–), J.W. Phillips and S.J. Parsons, who worked in a military research laboratory and mapped the entire northern sky at 64 MHz using a modified radar antenna and receiver.

Reber made the first suggestion that thermal radiation or *bremsstrahlung* might be the cause of the background emission, but this notion, despite its popularity, never could satisfactorily account for the strong evidence of a non-thermal spectrum. A rival theory favoured by A. Unsöld, M. Ryle, and G. Westerhout and J.H. Oort, proposed the existence of a population of hypothetical radio stars, each having an extremely high ratio of radio emission to optical emission compared with the Sun (see below), but this too ran into grave difficulties, primarily in accounting for the spatial distribution of the observed radiation. The first person to make the critical link between cosmic rays and the galactic background was K.O. Kiepenheuer (1910–75) who in 1950 briefly suggested that relativistic electrons spiralling in a general galactic magnetic field would emit synchrotron radiation matching the observed characteristics. In 1950 H.O.G. Alfvén and N. Herlofson also suggested the synchrotron mechanism (but in the context of a region the size of the solar system) as a model for the discrete 'radio stars' (see below). Although largely ignored in the West for many years, the synchrotron theory was enthusi-astically embraced and developed in the Soviet Union by V.L. Ginzburg, I.S. Shklovsky and their collaborators.

Solar radiation

Attempts prior to Jansky

The Sun was not only the first discrete source of extraterrestrial radio emission to be detected, but it was also the object of several unsuccessful experiments long before the time of Jansky. In the fifteen years following the laboratory experiments of Heinrich Hertz (1857–94), attempts to detect solar radio waves were made in several countries. In fact as early as 1890 Thomas A. Edison (1847–1931) and his associate A.E. Kennelly (1861–1939) are known to have considered exploring such a possibility through the use of an antenna consisting of a cable draped on poles around a mass of iron ore. The next known attempt was made in England sometime during the period 1897–1900 by Oliver J. Lodge (1851–1940), an early leader in the development of electrical techniques. Lodge's technique employed a copper cylinder housing a galvanometer, battery, and coherer as detector. The apparatus, including a short piece of wire as the entire extent of the antenna, was shielded from the Sun by a blackboard to filter out visible and infrared waves. In the only brief published comment on his results he simply remarks that he had great problems with interference caused by the electrical environment of urban Liverpool.

In 1896 J. Wilsing (1856–1943) and J. Scheiner (1858–1913), well known astrophysicists whose careers were primarily concerned with photographic and spectroscopic work on a variety of objects, carried out at Potsdam the most quantitative and detailed of the early experiments. Their apparatus, consisting of a resistive detector completely enclosed in a shielded box except for one small opening leading to a heliostat, was carefully calibrated and tested. They met with no success on any of eight days of observations, however, and finally concluded that they were being foiled by atmospheric absorption of the solar rays. This conclusion in turn prompted C. Nordmann (1881–1940) to make an attempt from a research station high in the French Alps. Despite what today appears to be a suitable antenna (fully 175 m long) and a receiver of adequate sensitivity for solar bursts, his one day of observations in Sep-

11.2. Grote Reber standing next to a 7.5-m Würzburg reflector operated by the US National Bureau of Standards at Sterling, Virginia in 1948. These high quality radar dishes were captured from the Germans at the close of World War II and played important roles in early radio astronomy in England, the Netherlands, the USA, France, and Sweden.

tember 1901, a time of solar minimum, yielded no signals.

Early discoveries

The intense development of the techniques of radar and of radio communication during World War II made it almost inevitable that radio emission from the Sun would be soon detected. G. C. Southworth and Reber independently measured solar radio radiation, in 1942 (9400 MHz) and 1943 (160 MHz) respectively. Neither was aware of the other's work for several years; nor did they know of the first discovery of solar radio waves, that by Hey in February 1942 (finally published in the open literature in 1945). The detection of radio emission from the Sun happened accidentally when British Army radar sets picked up excessive interference. Fearing a new jamming capability by the Germans, the Army ordered an investigation by Hey, its chief troubleshooter for such matters. Hey quickly deduced that in fact it was the *Sun* that was causing this 'jamming' and that the enhanced radio emission was probably associated with an exceptionally active sunspot group then on the central meridian of the Sun.

Physicists and engineers, highly trained in radio techniques and looking for suitable fields of research after World War II, were easily attracted to the exciting new field of solar radio astronomy in many countries including Australia, England, France and the United States. The radio equipment was almost exclusively surplus military radar and communications gear, much of which had been developed during the war by its new owners who now adapted it to astronomical purposes, for example, by changing antennas to equatorial mounts, eliminating the 'transmit' sections of the radar receivers and employing new tricks for increased receiver gain stability and sensitivity.

Post-war developments

An important early contribution by the radio workers to solar physics was the evidence provided by J. L. Pawsey (1908–62) in 1946 that the brightness temperature of the quiet Sun at metre wavelengths is $\approx 10^6$ K and the concomitant explanations by D. F. Martyn, Ginzburg and Shklovsky that this was a result of large radio opacity in a hot corona. The million-degree temperature of the solar corona had been previously suggested by

spectral and eclipse observations at visual wavelengths, but the radio data were perhaps the most convincing.

When better angular resolution was found to be necessary, the first solar radio observers, who typically were operating at frequencies between 50 and 200 MHz, extended radio techniques in ways familiar to optical astronomers. High resolution was achieved either by taking advantage of an eclipse or, more handily, by use of an interferometer. Among the eclipse expeditions were those of S. E. Khaikin and B. M. Chikhachev on a Soviet naval ship off the shore of Brazil in 1947 and the efforts on shorter wavelengths at partial eclipses by R. H. Dicke and R. Beringer in the United States in 1945, A. E. Covington in Canada in 1946, and W. N. Christiansen, D. E. Yabsley and B. Y. Mills in Australia in 1948.

At the same time groups at Cambridge University under Ryle and at the Radiophysics Laboratory in Sydney under E. G. Bowen and Pawsey developed the radio analogues to the Michelson interferometer and 'Lloyd's mirror'. Ryle and D. D. Vonberg used two aerials spaced at distances as large as 140 wavelengths to discriminate against the galactic background and allow more accurate studies of the Sun, and so were able to establish that solar radio bursts of short duration were often circularly polarized and of angular size less than 10 minutes of arc. They also developed a very effective balanced radiometer that was largely insensitive to receiver gain fluctuations. Simultaneously in Australia L. L. McCready, Pawsey and R. Payne-Scott showed that the radio bursts were small in angular size, correlated with solar active regions and very non-thermal in origin, having brightness temperatures often well above 10^8 K. Their technique involved a direct extension of their wartime experience in 'radio direction finding' for locating aircraft. Using an antenna already situated for defence on a seaside cliff, they could observe interference fringes between the direct wave from the Sun and the sea-reflected wave. This arrangement, analogous to Lloyd's mirror in optics, was used for some five years in Australia, but gradually the greater flexibility of the Michelson-type interferometer led to the almost universal adoption of the latter in the 1950s. Also in the late 1940s both the Cambridge and Sydney groups began designing experiments for mapping the (time-averaged)

11.3.(a) J. P. Wild adjusts one of the broadband rhombic antennas comprising his second dynamic spectograph (1952) at Dapto, New South Wales, Australia. (b) The setting circles of the rhombic antenna for Wild's original dynamic spectrograph (1950) at Penrith, New South Wales. Note that the tracking of the Sun was necessary to an accuracy of only a few degrees; this was typical of the poor angular resolution of early radio astronomy.

brightness distribution of the radio Sun at various wavelengths through the use of interferometers and Fourier techniques. For instance in 1950 H. M. Stanier argued from his measurements of the Sun's fringe visibility at interferometer spacings ranging up to 365 wavelengths that it exhibited no limb brightening at 60-cm wavelength.

A different approach was necessary to study temporal activity, however, and by 1950 much progress was being made, almost entirely through the efforts of J. P. Wild and McCready. These men overcame many technical problems and developed a radio dynamic spectrograph that allowed study of the spectrum of a burst with time resolution of less than a second over all frequencies in the 70 to 130 MHz range. These data allowed a systematic classification of the bursts into the now familiar Types I, II and III, which soon led to a physical understanding of the various emission mechanisms.

Discrete sources

The discrete radio sources were central to the development of radio astronomy from the time of their initial recognition. Although definite enhancements were noted at certain positions in Reber's 1944 map of the northern sky, the first true recognition of a discrete source came about in 1946 through the accidental discovery by Hey, Parsons and Phillips of rapid intensity fluctuations on a time scale of seconds from a region in the constellation of Cygnus. They found the 'source of disturbance', soon to be known as Cyg A, to be located in a region no larger than two degrees in extent. Assuming that the intensity variations were intrinsic to the source, they argued that it must in fact consist of a small number of discrete sources, each variable in a manner not unlike the Sun. Using the sea-cliff interferometer in Sydney, J. G. Bolton and G. J. Stanley soon thereafter established that Cyg A was in fact smaller than 8 minutes of arc in size. Arguments ensued, however, as to whether the radio scintillations were intrinsic to the source or were caused by irregularities in the Earth's ionosphere in a manner directly analogous to optical scintillation of stars in the troposphere. Finally, three studies in 1950, by C. G. Little and A. C. B. Lovell of Jodrell Bank, by F. G. Smith at Cambridge and by Stanley and O. B. Slee in Australia, established an ionospheric origin primarily by showing how the correlation between the fluctuations as observed at two stations decreased as the stations were further separated.

In 1948 Ryle and Smith discovered the strongest source of all, Cas A, and meanwhile Bolton and co-workers began to obtain positions for a few sources accurate to 10 minutes of arc. They employed the sea-cliff interferometric technique at a number of sites in Australia and New Zealand, and derived positions good enough in three cases to make identifications with objects known to optical astronomers. And these were no ordinary objects: the position of Tau A coincided with the Crab Nebula, Vir A with a bright elliptical galaxy (M 87) long known to have a peculiar jet in its nucleus, and Cen A with an unusual nebula, NGC 5128, bisected by a prominent dust band. But in general the first radio sources could not be unambiguously identified with anything on a photograph, simply because the positional errors were typically 1 to 3°. The number of radio sources discovered at Cambridge and Sydney nevertheless steadily grew – in 1950 Stanley and Slee compiled a catalogue of 22 southern radio sources while in the same year Ryle, Smith and B. Elsmore issued a catalogue (the first Cambridge or 1C survey) of 50 northern sources.

The Cambridge catalogue was made possible by a technical innovation of Ryle's whereby an interferometer could be made insensitive to the broad galactic background radiation as well as to man-made interference, thus allowing far more accurate measurements of the intensities, positions, sizes and polarizations either of discrete 'radio stars' (as the sources were coming to be known) or of solar bursts. The method involves alternately inserting and removing a half-wavelength of cable into one arm of an interferometer, so causing the fringe pattern to shift by one-half lobe at a relatively fast rate. Synchronous

detection of the difference between the outputs of the interferometer at the two positions of the phase switch then yields the desired fringes for small sources only. The phase switching technique found immediate and wide acceptance, for example in Mills's 1952 survey of radio sources.

The beginning of the merger of radio astronomy with traditional (optical) astronomy began when radio positions could be determined with an accuracy of 1 minute of arc (comparable to the best accuracy of Tycho Brahe's observations!), thus allowing the major optical telescopes to make detailed studies of regions of intense radio emission. Smith at Cambridge painstakingly analysed the sources of errors in interferometric positions and used two Würzburg dishes (Figure 11.2) to obtain about 1 minute of arc accuracy for the two strongest sources in the sky, which at that time had yet to be optically identified. The precision of these new positions enabled W. Baade and R. Minkowski in 1951 to use the 200-inch (508-cm) Palomar telescope to advantage. Cas A turned out to consist of a fascinating network of filaments, many exhibiting broad emission lines indicating extremely high internal velocities of $\gtrsim 2000$ km/s. Its origin was uncertain, however, since no pattern was then evident in the velocities or positions of the filaments, as one might expect, for example, in a supernova remnant. But it was Cyg A that yielded the most far-reaching implications, for it turned out to be an inconspicuous seventeenth-magnitude galaxy: although at a distance of over 100 million light years (as deduced from its redshift and a Hubble constant of 550 km/s/Mpc), this was the second strongest radio 'star' in the sky! It became clear that the radio sources scattered over the sky represented a probe into regions of the universe beyond the reach of even the largest optical telescopes.

The 21-cm hydrogen line

The field of spectroscopy has been in radio astronomy the notable exception to the rule that observers are always several paces ahead of the theorists. During 1944, in the German-occupied Netherlands, H. C. van de Hulst, a graduate student under the guidance of Oort, gave a seminar on the status of extraterrestrial radio observations, which he soon wrote up for publication. Van de Hulst drew from the eleven published articles on

11.4. H.I. Ewen in 1951 on the scaffolding beside the fixed horn antenna at Lyman Physics Laboratory, Harvard University. Here Ewen and Purcell first detected the galactic 21-cm radiation from neutral hydrogen.

the topic, but he went far beyond them in working out many new theoretical possibilities concerned with (1) 'free-free' or thermal radiation as the mechanism producing the galactic background (see above), (2) the expected line strengths and detectability of the radio transitions arising during recombination of electrons and protons in regions of ionized hydrogen (concluding that pressure broadening would make these lines unobservable) and (3) applications of radio data to various cosmological models. But the paper is best remembered for his calculations concerning the feasibility of detecting the 21-cm transition arising from the hyperfine splitting of the neutral hydrogen atom. Although neither the frequency nor the strength of the line had been measured in the laboratory, van de Hulst derived from available nuclear data a frequency of 1411 MHz and con-

cluded that the line should be observable if receivers could be made more sensitive and if the transition probability were at least 1×10^{-16} s^{-1}. Van de Hulst's cautious prediction was followed up only by Shklovsky who likewise worked out much of the physics of the expected hydrogen line emission, as well as of spectral lines from interstellar deuterium and the molecules OH and CH.

The 21-cm line was finally detected in 1951 by two groups who converged on the discovery from quite different paths. H.I. Ewen and E.M. Purcell were physicists at Harvard University hoping to employ the line as a probe of interstellar conditions. On the other hand Oort had never lost sight of his desire to use the 21-cm line to study the distant structure of our Galaxy and to this end had organized radio astronomy under difficult conditions in the post-war Netherlands. Despite this much earlier start, Oort and his colleagues C. A. Muller and van de Hulst were delayed by their limited technical background, as well as by an unfortunate fire which destroyed their entire receiver at one point. At the time of the eventual discovery at Harvard in the spring of 1951, van de Hulst chanced to be spending a sabbatical semester at Harvard and in fact was a catalyst to the success of both groups. To the Harvard group he fed astrophysical information on expected line widths, intensities and frequencies, while to the Dutch he passed technical hints, notably concerning the use of a modified Dicke-type comparison radiometer whereby deleterious gain fluctuations are minimized by subtracting the outputs from two adjacent frequencies continuously switched against each other. The discovery papers in *Nature* in 1951 (accompanied by news that Christiansen and J. V. Hindman in Australia had also detected the line) bear testimony to the different philosophies and styles of the two groups. Ewen and Purcell discussed the astrophysics of the line emission detected with their fixed horn antenna, while Muller

and Oort concentrated on the effects of galactic rotation on the line profiles as observed at several different galactic longitudes with their steerable 7.5-m Würzburg reflector (Figure 11.2). The measured Doppler shifts and line widths clearly revealed a differentially rotating, thin sheet of neutral hydrogen. This was to be the beginning of a decade of research in which 21-cm surveys supervised by Oort revealed many details of the dynamics and structure of the Milky Way.

Further reading

David O. Edge and Michael J. Mulkay, *Astronomy Transformed* (New York and London, 1976)

J.S. Hey, *The Evolution of Radio Astronomy* (London, 1973)

Bernard Lovell, *The Story of Jodrell Bank* (London, 1968)

Joseph L. Pawsey and Ronald N. Bracewell, *Radio Astronomy* (Oxford, 1955)

Iosef S. Shklovsky, *Cosmic Radio Waves* (Cambridge, Mass., 1960)

Woodruff T. Sullivan, III (ed.), *Classics in Radio Astronomy* (Dordrecht, Boston and London, 1982)

Woodruff T. Sullivan, III (ed.), *The Early Years of Radio Astronomy* (Cambridge, 1984)

The world's largest telescopes, 1850–1950

BARBARA L. WELTHER

For historical purposes it is often desirable to know which were the largest refractors and the largest reflectors at any given time. The following tables are designed so that the reader may establish the four largest telescopes of each type at any time within the century from 1850 to 1950. For the most part they list each telescope by the year in which it was originally set up and used for scientific observations. In some cases, however, the year may actually specify when the craftsmen finished the instrument.

The foundation of the tables presented here is the data in Dimitroff and Baker's *Telescopes and Accessories*. Their lists were first reordered according to date, and then divided into either two or three sections according to size. For the refractors, the first section tabulates only those instruments 15 inches (38 cm) and larger in diameter in the period 1850–80; the second, 20 inches (51 cm) and larger in 1881–1924; and the third, 24 inches (61 cm) and larger in 1925–50. For the reflectors, the first section tabulates only those 30 inches (76 cm) and larger in 1850–1924; and the second, 40 inches (102 cm) and larger in 1925–50. Therefore, the tables exclude such large telescopes as the 20-inch lens mounted at Stockholm in 1931 and the 36-inch (91-cm) mirrors mounted at Edinburgh and Greenwich in 1929.

To verify and supplement the data in *Telescopes and Accessories*, it was necessary to consult other sources. Unfortunately, these often contradict one another or omit some essential facts altogether. Furthermore, it was difficult to decide how to tabulate some instruments because they have been moved from one site to another, have been remounted or remodelled and changed size in the process, or have been broken or scrapped with no record kept of the event.

The discrepancies between sources occur prim-arily in the size given for the diameter of the lens or mirror. One problem results from conversion between the metric and English systems. The following tables give the diameter in both systems. Roman type denotes the system in which the instrument was crafted; and italic type, the system into which the dimension was converted. A lens or mirror is listed under the original size of the telescope even if the size changed because it was stopped down or put into a larger tube after the telescope was remounted, remodelled and/or relocated. Thus, although a 76-inch (193-cm) pyrex blank was made for the 74-inch (188-cm) David Dunlap telescope, the instrument is listed as 74 inches.

Other discrepancies in the literature occur with respect to dates. A telescope can be completed in the factory a year or more before its installation, and its actual use may come still later, leading to a variety of apparently contradictory dates. Furthermore, some sources include telescopes that were being built at the time of publication, but were not actually mounted then or perhaps ever. On the other hand, some lists may omit eligible telescopes that were either unknown to the compiler or dismantled at the time of tabulation. Occasionally, the same telescope will have two apparently independent listings because it changed ownership and/or location. A good example is the 15-inch (38-cm) refractor by Grubb that was at Dun Echt from 1873 to 1894 and at Edinburgh thereafter.

Most of the telescopes in the tables made scientific contributions. However, some seem to have been constructed only to claim the title of being the largest of their day. The Craig 24-inch lens of 1852, for instance, appears to surpass the twin 15-inch Merz refractors at Pulkovo and Harvard. In fact, the spherical aberration of the Craig lens made it practically useless and it was dismantled by 1858.

The Paris 15-inch refractor of 1854 also appears to rival those in Russia and America. However, the quality of its lens was so poor that it had to be stopped down considerably. Finally, poor weather precluded Buckingham from making significant observations with his 21-inch lens of 1862. Therefore, the Merz 15-inch refractors were neither truly rivalled nor surpassed until 1866 when the Chicago Astronomical Society set up the Clark 18.5-inch refractor. Similarly, the Paris 125-cm (49-inch) refractor of 1900 would appear to surpass the Yerkes 40-inch. However, its focal length of 57 m made it impossible to mount the telescope in a dome. So, after the Paris Universal Exhibition of 1900 for which it was especially conceived and constructed, it was dismantled, and therefore is not listed here.

In the tables, columns one and two give the diameter of the objective or mirror in inches and centimetres, respectively; roman type denotes the system in which the telescope was crafted; and italic, the other system into which the diameter was converted and appropriately rounded. When both columns are in italic, the instrument was originally specified in some other unit such as Prussian inches. Columns three and four list the date and place the telescope was originally mounted, respectively; in most instances the telescope is still there. Column five names the firm or craftsman who figured the lens or mirror, followed by the one who manufactured the mounting, if different. Column six names the institution or private party who originally owned the telescope, followed by any information of its subsequent owners and current location.

Further reading

Large refractors of the world, *The Observatory*, vol. 21 (1898), 239–41, 270–1

Principal astronomical observatories, arranged according to the size of refractor, in J. H. Willsey (compiler), *Harper's Book of Facts* (New York and London, 1898), 587

George Z. Dimitroff and James G. Baker, *Telescopes and Accessories* (Philadelphia, 1945), 280–91

H. P. Hollis, Large telescopes, *The Observatory*, vol. 37 (1914), 245–52

Henry C. King, *The History of the Telescope* (London and Cambridge, Mass., 1955)

Gerard P. Kuiper and Barbara M. Middlehurst (eds), *Telescopes* (Chicago, 1960; vol. 1 of the series *Stars and Stellar Systems*), 239–52

REFRACTORS 1850–1950

	Inches	cm	Date	Place	Craftsmen	Observatory
15 inches and larger	12	30	1833	Cambridge, England	Cauchoix	Cambridge University
	14	35	1834	Markree Castle, Sligo County, Ireland	Cauchoix	Colonel E. J. Cooper
	15	38	1839	St Petersburg [Leningrad, USSR]	Merz & Mahler	Pulkovo Observatory
	15	38	1847	Cambridge, Mass., USA	Merz & Mahler	Harvard College Observatory
	24	61	1852	Wandsworth Common, London, England	T. Slater, W. Gravatt	The Reverend John Craig; dismantled by 1858
	15	38	1854	Paris, France	Lerebours, Brunner	Paris Observatory; poor quality, stopped down for use
	21	53	1862	Walworth Common, London, England	Buckingham	J. Buckingham; moved to Edinburgh City Observatory, Calton Hill, 1898
	18.5	47	1866	Chicago, Ill., USA	Clark	Kept at University of Chicago by Chicago Astronomical Society; transferred to Northwestern University, 1889; deeded to Northwestern University, 1929
	15	38	1870	Tulse Hill, London, England	Grubb	On loan to Huggins by Royal Society; given to Cambridge University, Solar Physics Observatory, 1910
	25	64	1871	Gateshead-on-Tyne, England	Cooke & Sons	R.S. Newall; transferred to Cambridge University, Solar Physics Observatory, 1890; presented to National Observatory of Athens, 1956
	15	38	1873	Aberdeen, Scotland	Grubb	Dun Echt Observatory; given to Royal Observatory of Scotland, Aberdeen, 1894
	26	66	1873	Washington, DC, USA	Clark	US Naval Observatory; moved to Georgetown Heights and remounted by Warner & Swasey, 1893
	15	38	1877	Brussels, Belgium	Merz & Son	Royal Observatory
	15.5	39	1879	Madison, Wis., USA	Clark	University of Wisconsin, Washburn Observatory
	19	49	1879	Strassburg [Strasbourg], Germany	Merz, Repsold	Imperial Observatory; dismantled during World War I; renovated by National Observatory, University of Strasbourg, France, 1920s
	19	49	1879	Milan, Italy	Merz	Milan Observatory
	15	38	1880	Tacubaya, Mexico	Grubb	National Astronomical Observatory
	15	38	1880	Floirac, France	Merz, Gautier	Observatory of the University of Bordeaux
	16	41	1880	Rochester, New York, USA	Clark	Warner Observatory; moved to Mt Lowe, Calif., 1893

	inches	cm	year	Location	Maker	Observatory
20 inches and larger	22	56	1880	Sicily, Italy	Merz	Mt Etna Observatory
	20	52	*c.*1880	Turin, Italy	Porro	Francisco Porro
	27	69	1881	Vienna, Austria	Grubb	University Observatory
	23	58	1882	Princeton, NJ, USA	Clark	Princeton University Observatory
	26	66	1884	Charlottesville, Va., USA	Clark	Leander McCormick Observatory, University of Virginia
	29	73	1885	Paris, France	Martin	Paris Observatory
	30	76	1885	St Petersburg [Leningrad, USSR]	Clark, Repsold	Pulkovo Observatory
	30	76	1886	Nice, France	Henry Brothers, Gautier	Bischoffsheim Observatory, University of Paris
	36	91	1888	Mt Hamilton, Calif., USA	Clark, Warner & Swasey	Lick Observatory
	33	83	1889	Meudon, France	Henry Brothers, Gautier	Paris Observatory
	24	60	1890	Paris, France	Henry Brothers, Gautier	National Observatory
	24	62	1891	Meudon, France	Henry Brothers, Gautier	Paris Observatory
	20	51	1892	Philippine Islands	Merz	Manila Observatory
	24	61	1893	Cambridge, Mass., USA	Clark	Harvard College Observatory; moved to Southern Station, Arequipa, Peru, 1896; then to Bloemfontein, S. Africa, 1927
	28	71	1893	Greenwich, England	Grubb, Ransomes & Sims	Royal Greenwich Observatory
	20	51	1894	Denver, Colo., USA	Clark, Saegmüller	Chamberlin Observatory, University of Denver
	24	61	1896	Flagstaff, Ariz., USA	Clark	Lowell Observatory
	27	68	1896	Berlin-Treptow, Germany	Steinheil, Hoppe	Archenhold Observatory; refigured by Zeiss
	26	66	1897	Greenwich, England	Grubb	Royal Greenwich Observatory
	40	102	1897	Williams Bay, Wis., USA	Clark, Warner & Swasey	Yerkes Observatory
	24	61	1901	Cape of Good Hope, S. Africa	Grubb	Royal Observatory
	20	50	1901	Potsdam, Germany	Steinheil, Repsold	Astrophysical Observatory
	31	80	1901	Potsdam, Germany	Steinheil, Repsold	Astrophysical Observatory
	24	61	1902	Oxford, England	Grubb	Radcliffe Observatory; moved to London University Observatory, Mill Hill, 1931
	24	61	*c.*1905	Santiago, Chile	Grubb	National Observatory
	24	60	1908	Bergedorf, Germany	Steinheil, Repsold	Hamburg Observatory
	24	61	1911	Swarthmore, Pa., USA	Brashear, Warner & Swasey	Swarthmore College, Sproul Observatory
	26	65	1912	Berlin, Germany	Zeiss	Berlin-Babelsberg University Observatory, Neubabelsberg
	20	52	1913	Setif, Algeria	Schaer	Jarry-Desloges Observatory; (bought for private observatory in Celles by G. Fournier of Paris)
	26	65	*c.*1913	Belgrade, Yugoslavia	Zeiss	University of Belgrade, Astronomical Observatory
	30	76	1914	Pittsburgh, Pa., USA	Brashear, Warner & Swasey	Allegheny Observatory
	20	51	1917	Oakland, Calif., USA	Brashear, Warner & Swasey	Chabot Observatory
	20	51	1922	Middletown, Conn., USA	Clark, Warner & Swasey	Wesleyan University, Van Vleck Observatory
24 inches and larger	26	66	*c.*1925	Johannesburg, S. Africa	McDowell, local shops	Yale-Columbia Southern Station; moved to Mt Stromlo, Canberra
	26.5	67	1925	Johannesburg, S. Africa	Grubb	Union Observatory
	27	69	1927	Bloemfontein, S. Africa	McDowell, local shops	University of Michigan, Lamont-Hussey Observatory
	24	60	1928	Lembang, Java	Zeiss	Bosscha Observatory
	26	65	1930	Mitaka, Japan	Zeiss	Imperial University Observatory, Tokyo
	24	61	1931	Saltsjobaden, Sweden	Grubb-Parsons	Stockholm Observatory

REFLECTORS 1850–1950

	Inches	cm	Date	Place	Craftsmen	Observatory
30 inches and larger	36	91	1839	Parsonstown [Birr], Ireland	Parsons	William Parsons, third Earl of Rosse; remounted in 1874; dismantled in 1920s
	20	51	1845	Patricroft, Manchester, England	Nasmyth	James Nasmyth; moved to Penshurst, Kent; now in Kensington Science Museum, London
	72	183	1845	Parsonstown [Birr], Ireland	Parsons	William Parsons, third Earl of Rosse
	24	61	1846	Starfield, near Liverpool, England	Lassell	William Lassell; moved to Malta, 1852; brought back and set up in England
	48	122	1860	Valletta, Malta	Lassell	William Lassell; brought back to England; speculum destroyed by 1877
	31	80	1864	Marseille, France	Foucault, Eichens	Marseille Observatory
	48	122	1869	Melbourne, Australia	Grubb	Melbourne Observatory; mounting moved to Mt Stromlo, 1954
	28	71	1872	Hastings-on-Hudson, NY, USA	Draper	Henry Draper; moved to Cambridge, Mass., 1886; glass mirror in Historical Scientific Instruments Collection, Harvard University
	47	120	1875	Paris, France	Martin, Eichens, Gautier	Paris Observatory; moved to Haute Provence Observatory, refigured by Couder, remounted by Secrétan, 1943
	36	91	1879	Ealing, London, England	Calver, Common	A. A. Common; sold to Edward Crossley, Bermerside, Halifax, Yorkshire, 1885; presented to Lick Observatory, Mt Hamilton, Calif., 1895; remounted, 1902–5
	33	83	1887	Toulouse, France	Henry Brothers, Secrétan	Toulouse Observatory
	30	76	1888	Sidmouth, England	Common	Norman Lockyer; moved to South Kensington, London, 1912
	60	152	1889	Ealing, London, England	Common	A. A. Common; second mirror ground and installed, 1891; bought by Harvard College Observatory, Cambridge, Mass., 1904; dismantled 1933
	36	91	c.1889	South Kensington, London, England	Common	Solar Physics Observatory, South Kensington; moved with Solar Physics Observatory to Cambridge, 1911; returned to Science Museum, South Kensington, 1954

	39	100	1893	Meudon, France	Henry Brothers, Gautier	Paris Observatory
	33	84	1896	La Plata, Argentina	Gautier, Zeiss	La Plata Observatory
	30	76	1897	Greenwich, England	Common, Grubb-Parsons	Royal Greenwich Observatory
	37	94	1904	Santiago, Chile	Brashear	Lick Observatory Station; now Catholic University Observatory
	30	76	1905	Helwan, Egypt	Common, Reynolds	Khedival Observatory
	30	76	1906	Pittsburgh, Pa., USA	Brashear	Allegheny Observatory
	60	152	1908	Mt Wilson, Calif., USA	Ritchey, Pease	Mt Wilson Observatory
	44	112	1910	Flagstaff, Ariz., USA	Clark	Lowell Observatory
	37	94	1911	Ann Arbor, Mich., USA	Brashear	University of Michigan Observatory
	39	100	1913	Bergedorf, Germany	Zeiss	Hamburg Observatory
	39	100	1913	Uccle, Belgium	Zeiss	Royal Observatory
	39	100	1913	Mont Saleve, Switzerland	Schaer	Geneva Observatory; moved to Jungfraujoch in 1920s
	30	76	1916	Córdoba, Argentina	Perrine, Observatory shops	Córdoba Observatory
	100	254	1917	Pasadena, Calif., USA	Ritchey, Pease	Mt Wilson Observatory
	39	100	1918	Petit-Saconnex, Switzerland	Schaer	Geneva Observatory; moved to Geneva, 1922
	72	183	1918	Victoria, BC, Canada	Brashear, Warner & Swasey	Dominion Astrophysical Observatory
	36	91	1922	Tucson, Ariz., USA	Brashear, Warner & Swasey	Steward Observatory
40 inches and larger	39	100	1926	Merate, Como, Italy	Zeiss	Astronomical Observatory
	49	125	1927	Babelsberg, Germany	Zeiss	University Observatory
	40	102	1928	Simeis, Crimea, USSR	Grubb-Parsons	Pulkovo Observatory
	40	102	1931	Saltsjobaden, Sweden	Grubb-Parsons	Stockholm Observatory
	69	175	1932	Delaware, Ohio, USA	Fecker, Warner & Swasey	Perkins Observatory
	60	152	1933	Bloemfontein, S. Africa	Fecker	Harvard College Observatory
	74	188	1935	Richmond Hill, Ont., Canada	Grubb-Parsons	David Dunlap Observatory
	61	155	1937	Oak Ridge, Mass., USA	Fecker	Harvard College Observatory
	82	208	1939	Mt Locke, Texas, USA	Lundin, Warner & Swasey	McDonald Observatory
	60	152	1940	Córdoba, Argentina	Fecker, Warner & Swasey	Córdoba Observatory
	74	188	1948	Pretoria, S. Africa	Grubb-Parsons	Radcliffe Observatory
	200	508	1948	Palomar Mountain, Calif., USA	Brown, Westinghouse	Palomar Observatory

ILLUSTRATIONS: ACKNOWLEDGEMENTS AND SOURCES

1.1 H.E. Roscoe, *The Life and Experiences of Sir Henry Enfield Roscoe* (London, 1906).

1.2 Courtesy of Owen Gingerich.

1.3 G. Kirchhoff, *Untersuchungen über das Sonnenspectrum und die Spectren der chemischen Elemente*, pt. 2 (Berlin, 1863), fig. 1a.

1.4 Sir William and Lady Huggins (eds), *Publications of Sir William Huggins's Observatory*, vol. 2 (London, 1909).

1.5 P.A. Secchi, *Les Étoiles*, vol. 1 (Paris, 1879); composite of plates VII and VIII.

2.1 *Philosophical Transactions*, vol. 152 (1862), pt. 1, p. 363.

2.2 *Recueil de Mémoires, Rapports et Documents relatifs a l'Observation du Passage de Vénus sur le Soleil*, vol. 1, pt. 2 (Paris, 1876), fig. 4 on plate following p. 113.

2.3 US Naval Observatory.

2.4 Harvard College Observatory.

2.5 Agnes M. Clerke, *A Popular History of Astronomy during the Nineteenth Century* (2nd edition, Edinburgh, 1887), frontispiece.

2.6 E. Mouchez, *La Photographie Astronomique* (Paris, 1887); courtesy of R. Barthalot.

2.7 Laws Observatory, University of Missouri-Columbia.

2.8 *Scientific American*, cover, 15 October 1887; courtesy of J. Lankford.

2.9 *Scientific American*, cover, 29 October 1887; courtesy of J. Lankford.

3.1 J.A. Repsold, *Zur Geschichte der astronomischen Messwerkzeuge von 1830 bis um 1900*, vol. 2 (Leipzig, 1914), fig. 103.

3.2 J. Norman Lockyer, *Stargazing: Past and Present* (London, 1878), frontispiece.

3.3 *Engineering*, vol. 29, issue of 2 January 1880, foldout stubbed in after p. 6.

3.4 J.A. Repsold, *Zur Geschichte der astronomischen Messwerkzeuge von 1830 bis um 1900*, vol. 2 (Leipzig, 1914), fig. 72.

3.5 Yerkes Observatory, courtesy of L. Hobbs.

3.6 Harvard College Observatory.

3.7 *Memoirs of the Royal Astronomical Society*, vol. 36 (1866–67), plate XI.

3.8 *Memoirs of the Royal Astronomical Society*, vol. 50 (1890–91), p. 185.

4.1 S.P. Langley, *The New Astronomy* (Boston, 1888), fig. 16, p. 15.

4.2 J.N. Lockyer, *The Chemistry of the Sun* (London, 1887), fig. 80, p. 212.

4.3 *The Astrophysical Journal*, vol. 28 (1908), plate XVIII, courtesy of Yerkes Observatory, University of Chicago.

4.4 *Annales de l'Observatoire d'Astronomie Physique de Paris*, vol. 1 (1896), plate X.

4.5 *The Astrophysical Journal*, vol. 50 (1919), plate VI, courtesy of Yerkes Observatory, University of Chicago.

5.1 Smithsonian Astrophysical Observatory, after Helen Sawyer Hogg.

5.2 American Association of Variable Star Observers.

5.3 *Lick Observatory Bulletin* no. 300 (1917), plate III, fig. 2, courtesy of D. Osterbrock.

5.4 *Lick Observatory Bulletin* no. 232 (1913), fig. 1, p. 143.

5.5 *Harvard College Observatory Circular* no. 173 (1912), p. 3.

5.6 *The Astrophysical Journal*, vol. 32 (1910), p. 199.

6.1 *Nature*, vol. 94 (1915), p. 618.

6.2 *Annals of the Harvard College Observatory*, vol. 28 (1901), pt. 2.

6.3 E.C. Pickering Collection, Director's files, Harvard University Archives.

6.4 Courtesy of E. Lamla. Astronomische Institut, Universität Bonn.

6.5 *Publikationen des Astrophysikalischen Observatoriums zu Potsdam*, vol. 22 (1911), fig. 6, p. 29.

6.6 *Nature*, vol. 93 (1914), p. 252.

7.1 Royal Greenwich Observatory.

7.2 Paris Observatory, courtesy of J. Counil.

7.3 Portfolio of photographs, *Poulkovo Astronomical Observatory*, USSR (Harvard College Observatory).

7.4 Harvard College Observatory.

7.5 US Naval Observatory.

7.6 Lick Observatory.

7.7 Courtesy of D.B. Herrmann, Archenhold-Sternwarte Berlin-Treptow.

8.1 American Institute of Physics, Niels Bohr Library, courtesy of Owen Gingerich.

8.2 *The Astrophysical Journal*, vol. 14 (1901), p. 222, courtesy of Yerkes Observatory, University of Chicago.

8.3 Mt Wilson Observatory, courtesy of Owen Gingerich.

8.4 Mt Wilson and Las Campanas Observatory, courtesy of American Institute of Physics, Niels Bohr Library.

8.5 Mt Wilson Observatory, courtesy of Owen Gingerich.

8.6 Mt Wilson Observatory, courtesy of Owen Gingerich.

8.7 Dominion Astrophysical Observatory, Victoria, British Columbia.

8.8 Mt Wilson Observatory, courtesy of Owen Gingerich. Permission from Palomar Mountain Observatory.

8.9 Mt Wilson Observatory, courtesy of Owen Gingerich. Permission from Palomar Mountain Observatory.

8.10 Palomar Mountain Observatory, courtesy of Hansen Planetarium.

9.1 *Engineering*, vol. 12 (1890) p. 49. (Also in *Annals of the Cape Observatory*, vol. 7 (1896).)

9.2 Agnes M. Clerke, *A Popular History of Astronomy*

INDEX

Ii